Welcome to the **Science, Space, and Supernatural Phenomena Edition of Intelligent Minds**, the trivia series crafted for the intellectually curious, the scientifically inclined, and the explorers of the unknown!

In this edition, we venture beyond the earthly realm, diving deep into the mysteries of the universe, the marvels of scientific discovery, and the enigmatic occurrences that defy explanation. From the vast expanses of space to the intricate workings of the human mind, each question is an invitation to delve into the frontiers of knowledge and the unexplained.

Prepare to challenge your understanding of the cosmos, the forces that govern our existence, and the phenomena that captivate our imaginations. Whether you are a seasoned scientist, an aspiring astronomer, or a curious skeptic, these pages are filled with intriguing facts, unexpected connections, and thought-provoking questions that will expand your perspective.

So grab your telescope, don your lab coat, and open your mind to the wonders of the universe. Who knows what revelations await? Whether you're playing solo or competing with friends, each question is an opportunity to showcase your intellect and maybe uncover a new fascination.

Embark on this journey with us, and let's explore the extraordinary, the mysterious, and the awe-inspiring aspects of our world and beyond. Are you ready to prove your mettle and maybe even discover something new? Let the adventure begin!

"The important thing is not to stop questioning. Curiosity has its own reason for existing." — Albert Einstein

"To confine our attention to terrestrial matters would be to limit the human spirit." — Stephen Hawking

"The universe is full of magical things patiently waiting for our wits to grow sharper." — Eden Phillpotts

"Somewhere, something incredible is waiting to be known." — Carl Sagan

"The most beautiful thing we can experience is the mysterious. It is the source of all true art and science." — Albert Einstein

Table of Contents

Table of Contents	**3**
The Discovery of Penicillin	**6**
Dark Matter	**9**
The Invention of the Internet	**11**
1960s: The Seeds of Connectivity	11
1970s: Networking Takes Shape	11
1980s: The Internet Spreads Its Wings	11
1990s: The Internet Explodes	12
2000s: The Age of Expansion and Social Media	12
2010s: Internet of Things and Beyond	12
Trivia	13
Globalization and Industry 4.0	**15**
The Evolution of Industrial Revolutions	15
Space Exploration	**17**
Quiz - Space Exploration	18
The Naming of Planets	22
The Mystery of Crop Circles	28
Animals in Space	30
The Phenomenon of Aurora Borealis	32
Marie Curie - Fact File	34
Biology Quiz	36
The Enigma of Intuition	39
Heisenberg Uncertainty Principle	42
Astrophysics	43
Near-Death Experiences and Life After Death	45
Supersonic Shocks in Outer Space	46
The Placebo Effect	47
The Study Where Participants Got Drunk	49
Black Holes	50
Debunking Space Myths	52
Space in Cinema: Reality vs. Fiction	54

Famous inventors and their inventions	58
Human Genome Project	**60**
Type 1 Thinking: Intuitive and Fast	61
Type 2 Thinking: Analytical and Slow	61
Interaction Between Type 1 and Type 2 Thinking	62
Quantum Entanglement	63
Multiple choice Quiz on physics	65
True Or False - Physics	67
Cosmic Alchemy	69
Supernova Nucleosynthesis	70
Supernova Nucleosynthesis	71
Formation of Planets and Life	72
Isaac Newton	72
Carl Linnaeus	75
Charles Darwin	77
Galileo Galilei	80
The Wonders of the World	**83**
Ancient Wonders of the World	83
Medieval Wonders of the World:	84
New 7 Wonders of the World	85
Natural Wonders of the World:	86
Inventions Quiz	87
Karl Benz and the First Automobile	91
Space Radiation: Types, Sources, and Effects	93
Crazy Things Elon Musk Has Done	94
Gene Therapy for Genetic Disorders	**100**
What is Gene Therapy?	100
How Does It Work?	101
Applications and Success Stories	101
Challenges and Future Directions	102
Deep-Sea Biodiversity	103
Memory Formation and Retention	105
The Science of Sleep	107
Answers: The Science of Sleep	108
Albert Einstein	**110**

Albert Einstein's Journey from Germany to the United States	110
Albert Einstein's Theories	111
Did You Know? Crazy Facts About Albert Einstein	113
What is Nanotechnology?	**115**
The Evolution of 21st Century Private Space Companies	**117**
SpaceX (Elon Musk):	117
2. Blue Origin (Jeff Bezos):	118
3. Virgin Galactic (Richard Branson):	118
4. Rocket Lab (Peter Beck):	119
5. Sierra Nevada Corporation (Eren and Fatih Ozmen):	119
The Future of Space Travel	**120**
2040: The Dawn of Space Tourism and Lunar Exploration	120
2050: Permanent Lunar Bases and Expanding Space Tourism	121
2060: Human Settlements on Mars and Advanced Space Habitats	121
2070: Interplanetary Travel and Space-Based Economies	122
Crazy Facts About the Future of Space Travel	
Bonus Section	**122**
Top 5 Science-Focused Countries in the World	126
Small Countries with Interesting Science and Technology Projects	128
Time Travel in Theory and Popular Culture	130
UFO Phenomena and Government Disclosure	133
Synthetic Biology: Combining Biology and Engineering	138
Exoplanets and the Search for Extraterrestrial Life: Fact File	141
Psychokinesis: Mind-Over-Matter Phenomena	143
15 Crazy Facts About Psychokinesis	144
Near-Death Experiences (NDEs): Understanding Consciousness	146
Crazy Facts About Near-Death Experiences	147
Harry Houdini: The Famous Magician and Skeptic	149
Nikola Tesla: The Genius Inventor	152
Sally Ride: The First American Woman in Space	156
Interaction of Science and Geography	159
Gravitational Waves	**161**

The Discovery of Penicillin

The Fortuitous Fungus

The Accidental Discovery: In 1928, Alexander Fleming, a Scottish bacteriologist, returned from a vacation to find that a mold called Penicillium notatum had contaminated his Staphylococcus cultures. To his surprise, the mold was killing the bacteria around it.

Instead of dismissing the contaminated dish, Fleming studied the mold and extracted the substance that was killing the bacteria, naming it penicillin.

From Mold to Medicine

The Challenge of Purification: Initially, Fleming struggled to isolate penicillin in a form that could be used as medicine. It wasn't until the late 1930s that a team at Oxford University, including Howard Florey and Ernst Boris Chain, developed methods to purify and produce it in significant quantities.

Wartime Wonder Drug

The mass production of penicillin was ramped up during World War II. It became known as the "wonder drug" for its efficacy in treating bacterial infections in wounded soldiers, saving countless lives and revolutionizing the treatment of infections.

- Fleming chose the name "penicillin" simply based on the name of the mold that produced it, Penicillium notatum. It's like naming a superhero after the city they saved!
- The challenge of producing penicillin in useful quantities was so great that initially, the precious few milligrams of the drug were extracted from the urine of patients who had taken it.

Global Impact and Distribution:

- **International Collaboration:** The development and mass production of penicillin involved collaboration between scientists in the UK and the US. Notably, the United States War Production Board played a crucial role in organizing and scaling up production.
- **Saving Lives:** Penicillin's impact was immediate and profound. It dramatically reduced the death rate from bacterial infections and diseases such as pneumonia, syphilis, gonorrhea, and rheumatic fever. During World War II, penicillin saved the lives of countless soldiers and civilians.
- **Post-War Penicillin:** After the war, penicillin became widely available for civilian use. It ushered in the age of antibiotics, transforming medical practice and public health.

Challenges and Innovations:

- **Production Methods:** Early production methods were inefficient. The breakthrough came with the discovery that Penicillium chrysogenum, a different strain of the mold, produced much higher yields of penicillin. Scientists used this strain to improve production efficiency.
- **Industrial Scale Production:** Industrial-scale production involved deep-tank fermentation techniques, allowing for the production of large quantities of penicillin. Pfizer was one of the first companies to successfully produce penicillin on a massive scale using these methods.

Legacy and Recognition:

- **Nobel Prize:** In 1945, Alexander Fleming, Howard Florey, and Ernst Boris Chain were awarded the Nobel Prize in Physiology or Medicine for their work on penicillin.
- **Medical Advancements:** Penicillin paved the way for the development of other antibiotics, revolutionizing the field of medicine and drastically reducing the mortality rates associated with bacterial infections.

- **Public Awareness:** Fleming became a public figure, often emphasizing the importance of using antibiotics responsibly to avoid the development of resistance, a message that remains relevant today.

Penicillin Resistance:

- **Emergence of Resistance:** Shortly after penicillin was introduced, some bacteria began to develop resistance. This was first noted in the late 1940s, prompting ongoing research into new antibiotics and strategies to combat bacterial resistance.
- **Modern Challenges:** Antibiotic resistance remains a significant public health challenge. Research continues into developing new antibiotics and alternative treatments to address resistant bacterial strains.

Interesting Facts:

- **Fleming's Laboratory:** Alexander Fleming's laboratory at St. Mary's Hospital, London, where he discovered penicillin, is now a museum dedicated to his work and legacy.
- **Cultural Impact:** Penicillin has been featured in numerous films, books, and documentaries, highlighting its significance in medical history.

Dark Matter

Dark matter remains one of the most enigmatic components of the cosmos, eluding direct detection yet demonstrably exerting a massive influence on the structure and dynamics of galaxies. Though it does not emit, absorb, or reflect light, its gravitational effects are evident in the motion of stars within galaxies and the movements of galaxies in clusters.

> *"Invisible, yet strong enough to influence whole galaxies. That's dark matter, the universe's hidden secret."*

Cosmic Glue:

Dark matter makes up about 27% of the universe's mass-energy composition. It acts as the gravitational glue that holds galaxies and galaxy clusters together, preventing them from flying apart due to the speed at which they rotate.

Detection Methods:

Scientists study dark matter through indirect means such as gravitational lensing, where the presence of dark matter bends light from distant galaxies, or through the analysis of cosmic microwave background radiation which provides clues about the early universe's composition.

Candidate Particles:

Theories abound regarding the composition of dark matter, with candidates including Weakly Interacting Massive Particles (WIMPs), axions, and sterile neutrinos, although none have yet been directly observed.

Galactic Formation and Behavior:

Understanding dark matter is crucial for explaining how galaxies form and evolve. The distribution of dark matter influences the shape and structure of galaxies, guiding the visible matter into familiar forms.

Dark Matter Halos:

One of the intriguing aspects of dark matter is the concept of dark matter halos, which are vast, invisible regions surrounding galaxies. These halos are believed to contain the bulk of a galaxy's dark matter and play a critical role in maintaining its stability and structure. The gravitational influence of these halos extends far beyond the visible edges of galaxies, affecting their rotation curves and the movement of satellite galaxies.

Ongoing Research and Future Prospects:

The search for dark matter continues to be a major focus in astrophysics and cosmology. Large-scale experiments, such as those conducted at the Large Hadron Collider (LHC) and deep underground laboratories, aim to detect dark matter particles directly. Additionally, space-based observatories like the James Webb Space Telescope (JWST) are expected to provide new insights into the distribution and properties of dark matter. As our understanding of dark matter improves, it promises to unlock deeper mysteries about the universe's composition, formation, and ultimate fate.

The Invention of the Internet

Dive into the fascinating journey of the internet, from a concept in scientific papers to a global network that connects billions of people. This trivia page offers a quirky and detailed timeline that highlights key milestones in the creation and expansion of the World Wide Web.

1960s: The Seeds of Connectivity

1962: Visionary thinker J.C.R. Licklider of MIT introduces the idea of an "Intergalactic Network" in papers discussing the potential of interconnected computers.

1969: The birth of ARPANET, funded by the U.S. Department of Defense, which connected four university computers in California and Utah, marking the operational beginning of the internet.

1970s: Networking Takes Shape

1973: Vint Cerf and Bob Kahn develop the Transmission Control Protocol (TCP), which allows diverse computer networks to connect and communicate with each other. This protocol later becomes TCP/IP, the backbone of internet communication.

1976: Queen Elizabeth II sends an email from a British army base, becoming one of the first state leaders to use the network.

1980s: The Internet Spreads Its Wings

1983: ARPANET officially adopts TCP/IP, and the concept of a network of networks is realized.

1989: Tim Berners-Lee, a British scientist at CERN, proposes the World Wide Web as an internet-based hypermedia initiative for global information sharing, which would utilize browsers, HTML, and URLs.

"The Internet gave us access to everything; but it also gave everything access to us."

1990s: The Internet Explodes

1990: Tim Berners-Lee develops the first web browser and server, "WorldWideWeb.app".

1993: The release of the Mosaic web browser makes the World Wide Web accessible to the general public, sparking the internet boom.

1995: The term "surfing the internet" is coined, capturing the new leisure activity of browsing the vast expanse of the web.

2000s: The Age of Expansion and Social Media

2004: Facebook goes online, originally designed as a college network before becoming a global social media powerhouse.

2006: Google acquires YouTube, cementing video sharing as a core part of internet culture.

2010s: Internet of Things and Beyond

2010: The introduction of 4G wireless communication technology significantly improves mobile internet access.

2016: The number of devices connected to the internet surpasses the world's population, highlighting the pervasive reach of digital connectivity.

Trivia

- **Email Predates the World Wide Web:** The first email was sent by Ray Tomlinson in 1971, well before the web was created.
- **The First Item Sold Online:** The first secure online transaction was famously a CD of Sting's album "Ten Summoner's Tales" in 1994.
- **Wi-Fi Coffee Pots:** One of the first 'internet of things' devices was a coffee pot in Cambridge University's, which was connected to the internet for monitoring coffee levels.
- **First Ever Domain Name:** The first domain name ever registered was Symbolics.com on March 15, 1985. It was registered by Symbolics, a computer manufacturer based in Massachusetts.
- **First Social Media Site:** Many people consider SixDegrees.com, launched in 1997, to be the first true social media site.
- **Origins of the Hashtag:** The use of the hashtag (#) for categorizing content on social media was first proposed by Chris Messina in a tweet in 2007, revolutionizing how we organize and discover content online.
- **The Rise of Blogging:** The term "weblog" was coined by Jorn Barger in 1997, later shortened to "blog" by Peter Merholz in 1999, leading to the explosion of personal and professional blogs in the 2000s.
- **First Tweet:** The first tweet was sent by Twitter co-founder Jack Dorsey on March 21, 2006. It read: "just setting up my twttr."
- **Internet Shopping Milestone:** Amazon, founded in 1994 by Jeff Bezos, sold its first book online, "Fluid Concepts and Creative Analogies" by Douglas Hofstadter, in 1995, marking the start of e-commerce.
- **Birth of Wikipedia:** Wikipedia, the free online encyclopedia that anyone can edit, was launched on January 15, 2001 rapidly becoming a cornerstone of online information sharing.

"The internet is becoming the town square for the global village of tomorrow." — Bill Gates

Globalization and Industry 4.0

Globalization refers to the process by which businesses or other organizations develop international influence or start operating on an international scale. It's characterized by a freer flow of goods, services, money, people, and ideas across borders, facilitated by advances in transportation and telecommunications.

The Evolution of Industrial Revolutions

Industry 1.0:

Originating in the late 18th century, the First Industrial Revolution marked the transition from hand production methods to machines through the use of steam power and water power.

Industry 2.0:

The Second Industrial Revolution, starting in the late 19th century, introduced mass production with the help of electric power, creating a boom in productivity and the expansion of industries such as steel, oil, and electricity.

Industry 3.0:

Beginning in the 1970s, this phase brought about automated production through electronics and information technology, leading to more advanced manufacturing and data management systems.

Industry 4.0:

Now, Industry 4.0 represents the fourth revolution, which is characterized by a fusion of technologies that blur the lines between the physical, digital, and biological spheres. This phase integrates cyber-physical systems, the Internet of Things (IoT), and the Internet of Systems to make smart factories a reality. In these smart factories, machines are connected as a cohesive and intelligent network that can visualize the entire production chain and make decisions on its own.

A smart refrigerator can now keep track of your food's expiry dates, suggest recipes based on the contents, and even order groceries online when you're running low.

The rise of Industry 4.0 has led to the development of 'cyber twins,' which are replicas of physical systems that are kept in cyberspace to fool cyber attackers and protect actual critical data and operations.

Using IoT sensors and machine learning algorithms, predictive maintenance can forecast equipment failures before they happen, reducing downtime and maintenance costs significantly. (IoT: Internet of Things - network of connected devices; Machine Learning: a type of AI that learns from data)

AR is being used to train workers on complex machinery by overlaying digital information on real-world equipment, making training more intuitive and effective. (AR: Augmented Reality - technology that superimposes digital content on the real world)

Blockchain technology is being implemented to enhance supply chain transparency, allowing for real-time tracking of goods and verification of authenticity, reducing fraud and improving efficiency. (Blockchain: a decentralized ledger technology that records transactions across many computers)

Unlike traditional industrial robots, cobots are designed to work alongside human workers safely and efficiently, enhancing productivity and reducing the risk of injuries. (Cobots: robots designed for direct interaction with humans within a shared space)

Space Exploration

The 20th century heralded unprecedented advancements in space exploration, marking the era as a pivotal point in human history. The most iconic milestone occurred in 1969 when NASA's Apollo 11 mission successfully landed the first humans, Neil Armstrong and Buzz Aldrin, on the moon. This monumental achievement not only fulfilled President John F. Kennedy's pledge to land a man on the moon and return him safely to Earth within the decade but also symbolized the peak of the Space Race between the United States and the Soviet Union. This era of exploration was driven by both geopolitical rivalry and a boundless curiosity about the cosmos, leading to rapid developments in technology and our understanding of space.

The impacts of space exploration have extended far beyond the immediate scientific gains and national pride. Technologies developed during these missions have spun off into myriad applications that benefit everyday life on Earth. Innovations such as satellite technology have revolutionized communication, broadcasting, and navigation systems globally. Advances in materials science developed for space missions have found uses in various industries, including sports, medicine, and automotive. Even the integrated circuit, which became a fundamental building block for modern electronics, saw its development significantly accelerated by the demands of space technology. Moreover, the perspective of Earth from space fostered a new environmental awareness, highlighting the planet's fragility and interconnectivity, which spurred global movements towards environmental conservation.

HOUSTON, WE'VE HAD A PROBLEM HERE.

Quiz - Space Exploration

1. Who was the first human to orbit the Earth?

    A. Neil Armstrong
    B. Yuri Gagarin
    C. John Glenn

2. What was the primary objective of NASA's Apollo program?

    A. To explore the outer planets
    B. To land a man on the Moon and return him safely to Earth
    C. To send a man into orbit around Earth

3. Which country launched the first artificial satellite, Sputnik, into space?

    A. United States
    B. Soviet Union
    C. United Kingdom

4. What is the name of the first successfully deployed telescope in space?

    A. Hubble Space Telescope
    B. James Webb Space Telescope
    C. Spitzer Space Telescope

5. Which of the following innovations was a direct result of space exploration technologies?

    A. The Internet
    B. GPS (Global Positioning System)
    C. LED lightbulbs

6. Which mission first successfully landed humans on the moon?

    A. Apollo 11
    B. Apollo 13
    C. Apollo 17

7. What significant event happened during the Apollo 13 mission?

    A. It was the first to orbit the Moon
    B. It successfully landed two astronauts on the Moon
    C. It suffered a critical failure but returned safely to Earth

8. Which astronaut is famous for saying, "That's one small step for man, one giant leap for mankind"?

    A. Buzz Aldrin
    B. Michael Collins
    C. Neil Armstrong

9. Which space probe was the first to fly by Pluto and provide detailed images of its surface?

    A. Voyager 1
    B. New Horizons
    C. Pioneer 10

10. What was the first space station ever built?

    A. Mir
    B. Skylab
    C. Salyut 1

11. Who was the first woman to travel into space?

    A. Sally Ride
    B. Valentina Tereshkova
    C. Peggy Whitson

12. What was the primary mission of the Mars Rover "Curiosity"?

    A. To search for past or present life on Mars
    B. To study the climate and geology of Mars
    C. To collect samples and return them to Earth

13. What is the name of the international project that aims to return humans to the Moon by 2024?

    A. Artemis Program
    B. Orion Program

C. Constellation Program

14. Which planet was the first to be discovered using a telescope?

   A. Uranus
   B. Neptune
   C. Pluto

15. Which of the following spacecraft is known for its "Grand Tour" of the outer planets?

   A. Galileo
   B. Cassini
   C. Voyager 2

Answers

1- B) Yuri Gagarin

Yuri Gagarin's historic flight on April 12, 1961, aboard Vostok 1 made him an international hero and a symbol of Soviet space achievements. His call sign for the mission was "Cedar."

2- B) To land a man on the Moon and return him safely to Earth

The Apollo program, initiated in 1961, achieved its primary objective on July 20, 1969, with the Apollo 11 mission. The program included a total of 11 manned missions, six of which successfully landed on the Moon.

3- B) Soviet Union

Sputnik 1, launched on October 4, 1957, was the world's first artificial satellite. It transmitted radio pulses that could be received on Earth, marking the beginning of the space race.

4- A) Hubble Space Telescope

Launched on April 24, 1990, by the Space Shuttle Discovery, the Hubble Space Telescope has provided stunning images of distant galaxies, nebulae, and other astronomical phenomena, revolutionizing our understanding of the universe.

5- B) GPS (Global Positioning System)

The development of GPS was heavily influenced by space technology. The first satellite in the U.S. GPS system, Navstar 1, was launched in 1978, and the system became fully operational in 1993.

THAT'S ONE SMALL STEP FOR MAN, ONE GIANT LEAP FOR MANKIND.

6- A) Apollo 11

On July 20, 1969, Neil Armstrong and Buzz Aldrin became the first humans to walk on the Moon, while Michael Collins orbited above. Armstrong's famous words were broadcast to millions of viewers on Earth.

7- C) It suffered a critical failure but returned safely to Earth

Apollo 13, launched on April 11, 1970, experienced an oxygen tank explosion en route to the Moon. The mission's crew used the Lunar Module as a "lifeboat" and successfully returned to Earth on April 17, 1970.

8-C) Neil Armstrong

Neil Armstrong's words during the Apollo 11 Moon landing became one of the most iconic phrases in history. Armstrong and Aldrin spent about 21 hours on the lunar surface.

9- B) New Horizons

Launched on January 19, 2006, New Horizons conducted a flyby of Pluto on July 14, 2015, providing the first close-up images of the dwarf planet and its moons, revealing a diverse and complex world.

10- C) Salyut 1

Launched by the Soviet Union on April 19, 1971, Salyut 1 was the world's first space station. It was intended for scientific research and experimentation, but the mission ended in tragedy when the crew of Soyuz 11 perished during re-entry.

11- B) Valentina Tereshkova

Valentina Tereshkova flew into space on June 16, 1963, aboard Vostok 6. She remains the only woman to have been on a solo space mission, and her flight lasted almost three days.

12- B) To study the climate and geology of Mars

Launched on November 26, 2011, and landing on Mars on August 6, 2012, Curiosity's mission includes investigating Mars' habitability, studying its climate and geology, and collecting data for future human exploration.

13- A) Artemis Program

Named after Apollo's twin sister in Greek mythology, the Artemis program seeks to land the first woman and the next man on the Moon. It aims to establish sustainable exploration by the end of the decade.

14- A) Uranus

Discovered by Sir William Herschel on March 13, 1781, Uranus was the first planet found with a telescope. It was initially thought to be a comet or star before being recognized as a new planet.

15- C) Voyager 2

Launched on August 20, 1977, Voyager 2 is the only spacecraft to have visited all four gas giant planets: Jupiter, Saturn, Uranus, and Neptune. It continues to send data back to Earth from interstellar space.

THE STARS DON'T LOOK BIGGER, BUT THEY DO LOOK BRIGHTER.

The Naming of Planets

Mercury

Named by the Romans after the messenger god, Mercury.

Mercury, the closest planet to the Sun, was named by the Romans after the swift-footed messenger god, Mercury. This name was apt due to the planet's rapid movement across the sky, completing an orbit around the Sun in a mere 88 Earth days. The god Mercury, known as Hermes in Greek mythology, was renowned for his speed and agility, often depicted with winged sandals and a winged helmet. This god's attributes were seen as a perfect match for the fast-moving celestial body observed by ancient astronomers.

Reason and Meaning: Mercury was named for its speed and mobility, traits that the god Mercury exemplified through his roles in communication, travel, and commerce. The planet's swift orbit around the Sun makes it the fastest of all the planets, mirroring the god's legendary swiftness. This naming also reflects the planet's visibility during twilight hours, appearing briefly but brightly in the sky, akin to the fleeting yet impactful presence of the messenger god in myths. The choice of name underscores the importance of mythology in early astronomical studies, linking the physical characteristics of celestial bodies with human cultural narratives.

If our long-term survival is at stake, we have a basic responsibility to our species to venture to other worlds.

Venus

Named after the Roman goddess of love and beauty, Venus.

Venus, the second planet from the Sun, was named after the Roman goddess of love and beauty, reflecting its brilliant, shining presence in the sky. The goddess Venus, known as Aphrodite in Greek mythology, was associated with beauty, desire, and fertility. The planet's dazzling appearance, often the brightest object in the night sky after the Moon, made it a fitting tribute to the goddess who embodies radiance and allure. Ancient cultures were captivated by Venus's luminosity, leading to its association with one of their most revered deities.

Reason and Meaning: The planet Venus shines with a bright, steady light that has fascinated humans for millennia, often visible during dawn and dusk. This consistent and striking visibility made it a symbol of beauty and constancy, qualities attributed to the goddess Venus. In mythology, Venus's influence extended to love, attraction, and procreation, mirroring the planet's captivating and prominent presence in the heavens. The naming of Venus not only reflects its physical brightness but also highlights the cultural and symbolic significance attributed to it by ancient observers, linking the celestial with the divine in human imagination.

Earth

The name comes from Old English and Germanic words meaning "ground" or "soil."

Unlike the other planets in our solar system, Earth's name is not derived from Greco-Roman mythology. Instead, it originates from Old English and Germanic words like "eorðe" and "erde," which mean "ground" or "soil."

These terms emphasize the tangible, physical nature of our home planet, reflecting the human-centric view that has shaped our understanding of Earth throughout history.

Reason and Meaning: The name Earth underscores our connection to the land and the environment that sustains us. It is a reminder of our planet's life-sustaining properties, such as its fertile soil and abundant resources. The term reflects a practical and grounded perspective, contrasting with the mythological names of the other planets. It highlights our planet's unique role as the cradle of human civilization and life as we know it.

Mars

Named after the Roman god of war, Mars.

Mars, the fourth planet from the Sun, was named after the Roman god of war due to its reddish appearance. The ancient Romans associated the red color with blood and warfare, leading them to name the planet after Mars, their deity of war and aggression. In Greek mythology, this god is known as Ares, representing similar themes of conflict and strife.

Reason and Meaning: The reddish hue of Mars, caused by iron oxide (rust) on its surface, reminded ancient observers of blood, hence its association with the god of war. Mars symbolizes conflict, strength, and power, reflecting the harsh and barren landscape of the planet. This name also evokes the adventurous and often perilous nature of space exploration, as humanity strives to uncover the secrets of the Red Planet.

Jupiter

Named after the king of the Roman gods, Jupiter (Zeus in Greek mythology).

Jupiter, the largest planet in our solar system, was named after the king of the Roman gods, reflecting its massive size and dominant presence in the sky. The god Jupiter, known as Zeus in Greek mythology, ruled over the heavens and was associated with thunder, lightning, and authority. This naming highlights the planet's significance and grandeur.

Reason and Meaning: Jupiter's enormous size and strong gravitational influence make it a fitting namesake for the king of the gods. The planet's swirling clouds and powerful storms, including the Great Red Spot, mirror the mighty and tempestuous nature of the deity. Jupiter's name conveys a sense of power, authority, and majesty.

Saturn

Named after the Roman god of agriculture and time, Saturn.

Saturn, the sixth planet from the Sun, was named after the Roman god of agriculture, time, and wealth. This god, known as Cronus in Greek mythology, presided over the sowing and reaping of crops, reflecting themes of harvest and abundance. The planet's slow orbit and stately presence in the sky also align with the god's association with time.

Reason and Meaning: Saturn's name reflects the planet's slow and deliberate movement across the sky, completing an orbit around the Sun in about 29.5 Earth years. The planet's iconic rings, visible even with modest telescopes, add to its majestic and awe-inspiring appearance. Saturn symbolizes cycles, growth, and longevity.

Uranus

Named after the Greek god of the sky, Uranus.

Uranus, the seventh planet from the Sun, was named after the ancient Greek deity of the sky, Ouranos. This naming continued the tradition of using mythological figures, reflecting the planet's position in the heavens. Uranus was the primordial god of the sky, representing the overarching dome that encompasses the Earth.

Reason and Meaning: Discovered by William Herschel in 1781, Uranus was the first planet found with a telescope. Its name reflects its ethereal and distant nature, fitting for a planet located so far from Earth. Uranus's unusual tilt and blue-green color due to methane in its atmosphere add to its mysterious and otherworldly character, resonating with the vast and boundless sky.

Neptune

Named after the Roman god of the sea, Neptune.

Neptune, the eighth planet from the Sun, was named after the Roman god of the sea because of its deep blue color. The name was suggested due to the planet's appearance, which reminded astronomers of the ocean. Neptune, known as Poseidon in Greek mythology, ruled over the seas and was often depicted with a trident.

Reason and Meaning: The deep blue color of Neptune, caused by methane in its atmosphere, is reminiscent of the ocean's depths, making the name highly appropriate. Neptune's dynamic weather patterns, including the strongest winds in the solar system, reflect the turbulent and powerful nature of the sea god. This name emphasizes the planet's mysterious and majestic qualities, aligning it with the vast and unpredictable ocean.

Pluto (Dwarf Planet)

Named after the Roman god of the underworld, Pluto.

Pluto, discovered in 1930 by Clyde Tombaugh, was named after the Roman god of the underworld. The name was suggested by Venetia Burney, an 11-year-old girl from England. Pluto, known as Hades in Greek mythology, presided over the realm of the dead, fitting for a distant and cold celestial body.

Reason and Meaning: Pluto's remote location in the outer reaches of the solar system, along with its icy and dimly lit environment, made the name of the god of the underworld particularly fitting. Despite being reclassified as a dwarf planet in 2006, Pluto retains its mystique and continues to captivate scientists and the public alike. The name Pluto evokes themes of mystery, darkness, and the unknown, reflecting the planet's enigmatic nature and the challenges of exploring the farthest corners of our solar system.

- Pluto was reclassified from a planet to a dwarf planet in 2006 by the International Astronomical Union (IAU). This decision was based on the definition that a full-fledged planet must "clear its orbit" of other debris, a criterion Pluto does not meet.
- luto's surface is incredibly diverse, featuring mountains made of water ice, vast plains of frozen nitrogen, and possible cryovolcanoes.
- Pluto has an eccentric and tilted orbit, taking it as far as 49 astronomical units (AU) from the Sun and as close as 30 AU. Its orbit is also highly inclined compared to the planets of the solar system, intersecting the orbit of Neptune, which occasionally brings Pluto closer to the Sun than Neptune.

- NASA's New Horizons spacecraft made a historic flyby of Pluto on July 14, 2015.

The Mystery of Crop Circles

Embark on a journey into the enigmatic world of crop circles, those perplexing patterns that appear overnight in fields around the globe. These intricate designs range from simple circles to complex geometrical figures, creating a tapestry of mystery in the very grain of the countryside.

- **First Recorded Crop Circle:** The earliest recorded crop circle dates back to the 1600s. A woodcut from 1678, known as the "Mowing-Devil," shows a field of oat stalks laid out in a circle, with the devil cutting the oats.
- **Increase in Complexity:** Initially, crop circles were simple circles. Over time, they evolved into intricate patterns, some spanning over 200 meters in diameter, with detailed designs that have amazed both scientists and enthusiasts.
- **Alien Theories:** The appearance of crop circles near Stonehenge and other ancient sites fueled theories of extraterrestrial origins. Some UFO enthusiasts believe these formations are messages from alien beings or landing marks of UFOs.
- **Mathematical Marvels:** Many crop circles exhibit sophisticated mathematical properties, such as fractals and elements of sacred geometry, leading to speculation about their creation involving advanced knowledge.
- **Nighttime Creations:** Most crop circles appear overnight, adding to their mysterious nature. This has led to theories about them being created by otherworldly forces, as such complex designs would seemingly require more time and daylight to construct.

- **Cereologist:** The study and interpretation of crop circles is known as cereology. Enthusiasts who investigate crop circles are called cereologists.

Media Influence: The phenomenon of crop circles inspired various pop culture references, including movies like "Signs," which explore the alien theory behind these mysterious patterns.

Nocturnal Artists: Despite the confession of Doug and Dave, crop circle creators often work at night to maintain the mystery and avoid detection, using simple tools like ropes and boards to flatten crops.

Global Phenomenon: While crop circles are predominantly found in the UK, they have been reported in over 25 countries, with some of the other hotspots including the United States, Canada, Australia, and Japan.

The Doug and Dave Theory

In 1991, two men, Doug Bower and Dave Chorley, claimed they had been creating crop circles in England since the 1970s using planks, rope, and a baseball cap fitted with a loop of wire to help them walk in a straight line. Their confession demystified some of the formations but also added to the intrigue, as not all could be easily explained away.

Many crop circles exhibit a level of precision and complexity that baffles onlookers. Some designs are based on mathematical equations and sacred geometry, leading to speculations about their creation being beyond human capabilities.

The crop circle phenomenon has indeed permeated various aspects of culture, art, literature, and spirituality, often serving as a canvas for human creativity and curiosity. Here are specific examples of how crop circles have influenced different areas - Art, Literature, Spiritual Beliefs, ect.

- **Art:** Crop circles have inspired numerous artists. For instance, the intricate patterns have been replicated in large-scale sand sculptures by renowned sand artist Simon

Beck. Additionally, artists like Stan Herd have created 'earthworks' or large-scale land art that resembles crop circle designs.
- **Literature:** The mystery of crop circles has been explored in fiction. A notable example is "Signs" by M. Night Shyamalan, a novel adapted into a popular film that intertwines crop circles with a story of extraterrestrial contact. Another example is "The Circle Maker" by Mark Batterson, which, while not directly about crop circles, uses the concept metaphorically to explore themes of prayer and miracles.
- **Spiritual Beliefs:** Crop circles have given rise to new spiritual movements and beliefs. For some, these formations are seen as messages from higher powers or extraterrestrial beings. Books like "The Gift" by Freddy Silva explore crop circles as spiritual phenomena, suggesting they are created by non-human intelligence as messages to humanity. In some areas, they are woven into local legends, with stories of their creation ranging from whirlwinds and fairies to more extraterrestrial explanations.
- **Pop Culture:** Crop circles have been featured in television shows like "The X-Files," which often delved into paranormal phenomena, presenting crop circles as mysterious alien signatures.
- **Music:** The phenomenon has influenced music as well. For example, the electronic music duo Boards of Canada has an album titled "Geogaddi" with artwork and sounds inspired by crop circles, reflecting the mysterious and ambient nature of the phenomenon.

Animals in Space

The history of animals in space is not just a tale of human ingenuity, but a testament to the bravery of some unlikely astronauts. Long before humans set foot on the Moon, our furry and feathered friends were the pioneering space travelers, helping us understand if life beyond our planet was possible.

The Canine Cosmonauts: The Soviet space program sent the first canine cosmonauts, including the famous Laika, the first animal to orbit Earth in 1957. Though Laika's mission was one-way, she paved the way for human spaceflight.

Buzzing into Orbit: Believe it or not, fruit flies were the first Earthlings to reach space. In 1947, they were launched aboard a U.S. V-2 rocket to study radiation exposure at high altitudes.

Monkeys and Apes: The United States sent several monkeys and apes into space in the 1950s and 60s to study the biological effects of space travel. Ham, a chimpanzee, became famous in 1961 for his suborbital flight, demonstrating tasks during the flight before safely returning to Earth.

Félicette, the Astrocat: In 1963, France sent Félicette, a stray Parisian cat, into space. Félicette returned safely after a brief suborbital flight, making her the first and only feline astronaut to date.

Spacefaring Frogs and Newts: The amphibious set has also seen its day in orbit. NASA's Orbiting Frog Otolith experiment in 1970 involved sending frogs to space to study weightlessness's effects on the inner ear.

Spiders Spinning in Zero Gravity: In 1973, two garden spiders named Arabella and Anita lived aboard Skylab, spinning webs in space so scientists could study how zero-gravity affected their abilities.

Mice, Fish, and Even Jellyfish: Modern space missions have seen a diverse range of animals, including mice, fish, and jellyfish, helping researchers understand long-term effects of space on living organisms.

A Celestial Menagerie: The International Space Station has become a floating laboratory for studying various organisms, including worms and tardigrades, known for their resilience in extreme conditions.

The legacy of these animal astronauts lives on, not only in the advancements they contributed to space exploration but also in the ethical discussions they inspire about our responsibilities to our fellow Earthlings.

The Phenomenon of Aurora Borealis

The Aurora Borealis, also known as the Northern Lights, presents one of nature's most spectacular and enchanting displays. This luminous phenomenon, with its mesmerizing swirls of green, purple, and pink, lights up the polar skies, capturing the imagination of anyone fortunate enough to witness it. Here's a glimpse into the science and lore behind this celestial spectacle.

The Aurora Borealis is a result of interactions between the Earth's atmosphere and charged particles from the sun. These particles are carried towards the poles by the Earth's magnetic field and collide with gases like oxygen and nitrogen in the atmosphere, creating the dazzling light show.

The intensity and frequency of the Northern Lights are closely linked to solar activity. Solar flares and coronal mass ejections from the sun increase the number of charged particles interacting with the Earth's atmosphere, often leading to more vivid displays.

The varying colors of the aurora are due to the type of gas particles involved in the collisions. Oxygen emits green and red light, while nitrogen produces blue and purple hues. The blending of these colors creates the aurora's signature ethereal glow.

The Aurora Borealis, or Northern Lights, is a celestial spectacle that draws travelers to the far reaches of the northern hemisphere. Here's a guide to some of the best locations to witness this natural wonder and what you can do while you're there:

Tromsø, Norway: Often called the 'Capital of the Arctic', Tromsø offers a high chance of Northern Lights sightings from September to April. Visitors can combine aurora hunting with dog sledding, snowshoeing, and visiting the iconic Arctic Cathedral.

Reykjavik, Iceland: While you can catch the Northern Lights from the city, heading out to less urban areas like Thingvellir National Park increases your chances. Reykjavik also offers geothermal spas like the Blue Lagoon for a relaxing experience under the auroral glow.

Fairbanks, Alaska: Positioned under the 'Auroral Oval', Fairbanks is an ideal spot for viewing the lights. The city offers guided aurora tours and the chance to explore the stunning Alaskan wilderness by day, including visits to Chena Hot Springs.

Yellowknife, Canada: This city in Canada's Northwest Territories is known for its clear skies, making it perfect for Northern Lights viewing. Activities include ice fishing, snowmobiling, and learning about the indigenous cultures at the Prince of Wales Northern Heritage Centre.

Abisko, Sweden: Located in the Swedish Lapland, Abisko is almost free from light pollution and cloud cover, providing excellent aurora viewing conditions. Visitors can also enjoy a trip on the Aurora Sky Station and explore the picturesque landscapes of the Abisko National Park.

Rovaniemi, Finland: Known as the official hometown of Santa Claus, Rovaniemi offers a magical winter experience with aurora viewing. Enjoy reindeer sleigh rides, visit the Santa Claus Village, and stay in unique accommodations like glass igloos.

Luosto, Finland: This small resort town is home to the Aurora Chalet, where you can receive 'Aurora Alarms' when the Northern Lights appear. During the day, explore the Amethyst Mine or enjoy skiing and snowboarding.

Marie Curie - Fact File

- Marie Curie, born Maria Sklodowska, on November 7, 1867, in Warsaw, Poland, was a pioneering physicist and chemist.
- Curie moved to Paris to study at the Sorbonne, where she earned degrees in physics and mathematics, becoming one of the first women to achieve such accolades.
- Along with her husband, Pierre Curie, Marie discovered the elements polonium and radium, significantly advancing the understanding of radioactivity, a term she coined.
- Marie Curie was the first woman to win a Nobel Prize and the only person to win Nobel Prizes in two different scientific fields: Physics (1903) and Chemistry (1911).
- She established the Radium Institute in Paris, which became a leading center for nuclear physics and chemistry research.
- During World War I, Curie developed mobile radiography units, known as "Little Curies," to assist in the medical treatment of wounded soldiers on the front lines.
- Despite facing significant challenges as a woman in science, Curie's perseverance led to groundbreaking discoveries, though her exposure to high levels of radiation likely contributed to her later health issues.

"Nothing in life is to be feared, it is only to be understood" - Marie Curie

- Her daughter, Irène Joliot-Curie, also won a Nobel Prize in Chemistry, continuing the family legacy in scientific achievement.
- Marie Curie died on July 4, 1934, from aplastic anemia, a condition linked to prolonged radiation exposure. She was later interred in the Panthéon in Paris, honoring her significant contributions to science and humanity.

- Marie Curie's research laid the groundwork for the development of X-ray machines and cancer treatments, and she remains an iconic figure in science and medicine.
- Marie Curie's early education was largely self-directed; she studied in secret as part of the "Flying University," an underground educational initiative in Poland that provided opportunities for women when they were banned from higher education.
- Curie was the first female professor at the University of Paris (Sorbonne), where she took over her late husband's professorship, breaking significant gender barriers in academia.
- Curie's personal notebooks and research papers are still highly radioactive and are stored in lead-lined boxes; they remain too dangerous to handle without protective equipment, illustrating the lasting impact of her groundbreaking work on radioactivity.
- Marie Curie's legacy extends to numerous institutions and awards named in her honor, including the Marie Curie Cancer Care organization in the UK and the Curie Institute in Paris, which continues to be a leading research center in medical science.

"The universe is not only stranger than we imagine, it is stranger than we can imagine." — Arthur Eddington

"There are more things in heaven and earth, Horatio, than are dreamt of in your philosophy." — William Shakespeare, Hamlet

"I do believe in an everyday sort of magic—the inexplicable connectedness we sometimes experience with places, people, works of art, and the like; the eerie appropriateness of moments of synchronicity;

the whispered voice, the hidden presence when we think we're alone." — Charles de Lint

"Mystery creates wonder and wonder is the basis of man's desire to understand." — Neil Armstrong

"The oldest and strongest emotion of mankind is fear, and the oldest and strongest kind of fear is fear of the unknown." — H.P. Lovecraft

Biology Quiz

1. What is the basic unit of life?

    A. Atom
    B. Cell
    C. Molecule

2. Which of the following is the process by which plants make their own food?

    A. Respiration
    B. Photosynthesis
    C. Fermentation

3. What molecule carries genetic information in most living organisms?

    A. RNA
    B. DNA
    C. Protein

4. Which organ in the human body is primarily responsible for filtering blood?

    A. Liver
    B. Heart
    C. Kidney

5. What is the powerhouse of the cell?

   A. Nucleus
   B. Mitochondrion
   C. Ribosome

6. Which of these is NOT a type of carbohydrate?

   A. Glucose
   B. Starch
   C. Protein

7. What type of macromolecule are enzymes?

   A. Lipids
   B. Proteins
   C. Nucleic acids

8. Which blood cells are responsible for fighting infections?

   A. Red blood cells
   B. White blood cells
   C. Platelets

9. What term describes the variety of all the genes, species, and ecosystems in a given place?

   A. Biodiversity
   B. Ecology
   C. Evolution

10. Which of the following best describes a gene?

    A. A segment of DNA that codes for a protein
    B. A type of cell
    C. A molecule that provides energy for cellular processes

True or False: Humans have more bacterial cells in their body than human cells.

True or False: Photosynthesis only occurs in plants.

True or False: Mitochondria are often referred to as the "powerhouse of the cell."

True or False: All cells in the human body contain the same DNA.

True or False: Evolution is a theory that explains the diversity of life on Earth.

Did You Know?

Octopuses have three hearts! Two pump blood to the gills, while the third pumps it to the rest of the body. Additionally, their blood is blue because it contains a copper-based molecule called hemocyanin, which is more efficient at transporting oxygen in cold, low-oxygen environments than the iron-based hemoglobin found in human blood.

Biology Quiz Answers

1. B) Cell
2. B) Photosynthesis
3. B) DNA
4. C) Kidney
5. B) Mitochondrion
6. C) Protein
7. B) Proteins
8. B) White blood cells
9. A) Biodiversity
10. A) A segment of DNA that codes for a protein

True or False Answers

1. True
2. False
3. True
4. True
5. True

The Enigma of Intuition

Whether referred to as gut feelings, a 'sixth sense,' or another term, intuition is a universal human experience. We've all encountered moments where we just "know" something without conscious reasoning. These instincts, while sometimes inaccurate—like the unsettling certainty felt during airplane turbulence—often prove surprisingly accurate. Psychologists suggest that our brains constantly collect and process vast amounts of information from our surroundings. This subconscious data processing can lead to moments where we sense or understand something without being able to articulate how or why. These intuitive moments hint at a complex interplay between our conscious and unconscious minds. However, the mysterious nature of intuition makes it challenging to study and prove scientifically. While psychology provides some insights, it might only scratch the surface of understanding this phenomenon. Intuition could involve a blend of cognitive processes, emotional cues, and even evolutionary mechanisms that have developed to enhance survival. The subjective nature of intuition, combined with its sometimes uncanny accuracy, continues to intrigue researchers and the general public alike. As we explore the boundaries of human cognition, the enigma of intuition remains a captivating frontier.

"Intuition is a very powerful thing, more powerful than intellect, in my opinion." — Steve Jobs

One intriguing study on intuition was conducted by researchers from Tel Aviv University and published in the journal Psychological Science in 2017. The study aimed to investigate the validity and mechanisms behind intuitive decision-making. Participants were asked to perform a series of complex decision-making tasks, which were designed to be too difficult to solve using conscious reasoning alone. Interestingly, the study found that participants who relied on their intuition performed better than those who attempted to analyze the tasks logically.

In one part of the experiment, participants were shown a rapid sequence of images and were asked to choose the "winning" image based solely on their gut feeling. Despite the speed and complexity of the task, many participants were able to consistently identify the correct images. This suggests that intuition can sometimes access patterns and insights that the conscious mind cannot easily discern.

The researchers concluded that intuition might stem from the brain's ability to unconsciously process vast amounts of information, identifying patterns and making connections that are not immediately apparent to our conscious mind. This study provides compelling evidence that intuition is not merely a mystical or unreliable sense, but rather a sophisticated cognitive process that can enhance decision-making in certain contexts.

Did you know? Studies have shown that people who trust their intuition often make quicker decisions and are generally more confident in their choices, even in high-pressure situations. This confidence can sometimes translate into better performance, especially in dynamic and fast-paced environments.

Déjà Vu

"Déjà vu is the feeling that one has lived through the present situation before."

Déjà vu, a French term meaning "already seen," is a peculiar sensation experienced by many people at least once in their lifetime. It is the eerie feeling that a current situation is familiar, as if it has been lived through before, despite knowing that it is happening for the first time. This phenomenon is a common and often fleeting experience that leaves individuals questioning the nature of their memory and perception. Psychologists suggest that déjà vu may occur when the brain's memory systems momentarily malfunction, causing the present to be mistakenly perceived as a past memory. This misfire creates a sense of familiarity without a clear source, blending the boundaries between past and present experiences.

While déjà vu is widely recognized, it remains a mysterious and elusive phenomenon that is challenging to study in a controlled environment. Various theories attempt to explain its occurrence, ranging from neurological explanations involving temporal lobe activity to psychological interpretations that link it to subconscious recognition of familiar elements within a new context. Despite its enigmatic nature, déjà vu continues to intrigue scientists and laypeople alike, prompting ongoing research into the mechanisms behind this curious experience.

> *"Memory is the diary that we all carry about with us." — Oscar Wilde*

Fascinating Study on Déjà Vu

One notable study on déjà vu was conducted by researchers from Colorado State University and published in Psychological Science in 2006. The study aimed to explore the neural basis of déjà vu using virtual reality (VR) technology to recreate experiences that might trigger the sensation. Participants navigated through VR environments that were designed to be subtly similar to previous environments they had encountered, without being identical.

The researchers found that participants frequently reported experiencing déjà vu when they encountered new VR environments that contained familiar elements from previous ones. This supported the idea that déjà vu may arise from the brain recognizing patterns and similarities between new and past experiences, even if the conscious mind does not explicitly remember the details. The sensation of familiarity without clear memory recall points to complex interactions between memory systems in the brain. The study concluded that déjà vu might be linked to the brain's ability to detect and process similarities between current and past experiences, even when the specific memories are not consciously accessible. This research provides valuable insights into the cognitive processes underlying déjà vu and highlights the potential for VR as a tool in studying complex psychological phenomena.

Did you know?

Some researchers believe that déjà vu could be a form of minor seizure activity in the brain, specifically in the temporal lobe, which is responsible for processing memories and emotions

Heisenberg Uncertainty Principle

"Uncertainty is not a bug of quantum mechanics; it's a feature."

The Heisenberg Uncertainty Principle is a fundamental concept in quantum mechanics that was formulated by Werner Heisenberg in 1927. This principle challenges our traditional ideas about the predictability and determinism of the physical world.

The principle states that it is impossible to simultaneously know both the exact position and exact momentum of a particle. The more precisely one property is measured, the less precisely the other can be controlled or determined. The principle is often expressed in terms of inequalities, one of the most famous being:

$\Delta x * \Delta p \geq \hbar/2$

where Δx is the uncertainty in position, Δp is the uncertainty in momentum, and $\hbar$ (h-bar) is the reduced Planck's constant, which is a very small number indicating that the effect becomes significant only at the quantum level.

- Heisenberg formulated this principle while working at Niels Bohr's institute in Copenhagen, a hub for the early development of quantum mechanics.
- Werner Heisenberg was awarded the Nobel Prize in Physics in 1932 for the creation of quantum mechanics, the

application of which led, among other things, to the discovery of the allotropic forms of hydrogen.
- in fast-moving financial markets, traders must often make quick decisions with incomplete information.

Astrophysics

Astrophysics is a branch of astronomy that focuses on the physical properties and processes of celestial objects and phenomena. It combines principles of physics and astronomy to understand the behavior, composition, and evolution of stars, galaxies, planets, black holes, and other cosmic entities. Astrophysicists use mathematical models, computer simulations, and observational data to study a wide range of phenomena, from the birth of stars in distant galaxies to the behavior of matter in extreme environments like black holes. By probing the fundamental laws of physics at work in the cosmos, astrophysics sheds light on the origins and nature of the universe itself.

Cosmic Recycling

- Stars are born from vast clouds of gas and dust, condensing under their own gravity to ignite nuclear fusion in their cores.
- Throughout their lives, stars fuse lighter elements like hydrogen and helium into heavier elements such as carbon, oxygen, and iron.
- When massive stars exhaust their nuclear fuel, they undergo supernova explosions, scattering newly formed elements back into space to enrich future generations of stars and planetary systems.

Neutron Stars

- Neutron stars are incredibly dense stellar remnants left behind after massive stars undergo supernova explosions.

- Despite their small size, neutron stars pack an immense amount of mass, leading to extreme densities where a teaspoonful of material could weigh billions of tons.
- Neutron stars exhibit fascinating properties such as rapid rotation, powerful magnetic fields, and the potential to form exotic states of matter.

Black Hole Ballet

- Black holes are regions of spacetime where gravity is so intense that nothing, not even light, can escape from them.
- As matter falls into a black hole's gravitational grasp, it forms a swirling disk of gas and dust known as an accretion disk, emitting intense radiation across the electromagnetic spectrum.
- Some black holes also produce powerful jets of particles that shoot out from their poles, generating energetic phenomena observed across the universe.

Cosmic Microwave Background

- The cosmic microwave background (CMB) radiation is the residual heat leftover from the Big Bang, permeating the entire universe. Detected in 1965, the CMB provides a snapshot of the universe's state roughly 380,000 years after the Big Bang, when neutral atoms formed and photons began to travel freely. Studying the CMB allows astronomers to investigate the universe's early conditions, including its temperature variations and overall structure, offering vital clues about its origin and evolution.
- The discovery of the CMB radiation by Arno Penzias and Robert Wilson in 1965 earned them the Nobel Prize in Physics in 1978, confirming the Big Bang theory's predictions and revolutionizing cosmology. Detailed maps of the CMB, such as those generated by the Wilkinson Microwave Anisotropy Probe (WMAP) and the Planck

satellite, have provided precise measurements of the universe's age, composition, and expansion rate, leading to significant advancements in our understanding of cosmology.

Near-Death Experiences and Life After Death

Near-death experiences (NDEs) are profound psychological events that occur in individuals who come close to death or are in situations of extreme physical or emotional danger. These experiences often include sensations such as moving through a tunnel, encountering a bright light, feeling a sense of peace, and sometimes meeting deceased loved ones or spiritual beings. NDEs are reported across different cultures and age groups, leading to a widespread interest in understanding their nature and implications.

The phenomenon of NDEs raises compelling questions about the nature of consciousness and the possibility of life after death. While some interpret NDEs as evidence of an afterlife, others propose that these experiences are the result of physiological and psychological processes in the brain under extreme stress. The brain's response to oxygen deprivation, release of endorphins, or changes in brain chemistry during life-threatening situations are among the scientific explanations suggested. Despite extensive research, NDEs remain a fascinating and controversial topic, blending elements of spirituality, neurology, and psychology.

"Death is a challenge. It tells us not to waste time... It tells us to tell each other right now that we love each other." — Leo Buscaglia

Fascinating Study on Near-Death Experiences

One of the most comprehensive studies on near-death experiences was conducted by Dr. Pim van Lommel, a Dutch cardiologist, and published in The Lancet in 2001. This study aimed to investigate the nature of NDEs in cardiac arrest patients who were successfully resuscitated. Over a period of several years, Dr. van Lommel and his team interviewed 344 patients who had experienced cardiac arrest, documenting their accounts of NDEs.

The study found that about 18% of the patients reported having a near-death experience, characterized by elements such as out-of-body experiences, feelings of peace, seeing a bright light, and encountering deceased relatives. Interestingly, the study noted that these experiences were not influenced by the patients' religious beliefs, age, or cultural background. The consistency of NDE reports among diverse populations suggested that these experiences might stem from universal brain mechanisms activated during life-threatening events.

Dr. van Lommel's research concluded that NDEs could not be fully explained by current scientific knowledge of brain function and consciousness. The study sparked ongoing debate and further research into whether NDEs provide evidence of consciousness existing independently of brain activity.

Did you know?

Some studies suggest that the life review phenomenon reported during near-death experiences, where individuals see a rapid playback of their life's events, could be linked to the brain's attempt to process vast amounts of information in a short time during extreme stress.

Supersonic Shocks in Outer Space

In the vast expanses of outer space, supersonic shocks are extraordinary phenomena that occur when objects travel faster than the speed of sound in a given medium, creating powerful shock waves. These shock waves are similar to the sonic booms experienced on Earth when aircraft exceed the speed of sound, but in space, they can be caused by various high-energy events and interactions. Supersonic shocks play a crucial role in shaping the dynamics of the cosmos, influencing everything from the formation of stars to the behavior of interstellar gas.

- Formation in Supernovae: Supersonic shocks are commonly generated during supernova explosions, where the outer layers of a dying star are expelled at incredibly high speeds. The shock waves from these explosions compress and heat the surrounding interstellar medium, often triggering the formation of new stars.
- Solar Wind Shocks: The solar wind, a stream of charged particles emitted by the Sun, can create supersonic shocks when it collides with planetary magnetospheres or other obstacles in space. These shocks influence space weather and can affect satellite operations and communication systems on Earth.
- Bow Shocks Around Stars: Stars moving through the interstellar medium at high velocities can create bow shocks, similar to the bow wave formed by a boat moving through water. These shocks are visible as arcs of glowing gas and dust, often observed in infrared and radio wavelengths.

The Placebo Effect

"Belief is half of all healing." Plautus

The placebo effect is a fascinating psychological phenomenon where patients experience real improvements in their health or symptoms after receiving a treatment that has no therapeutic effect. This effect occurs when a person believes they are receiving a real treatment, leading to measurable changes in their condition despite the treatment being inactive, such as a sugar pill or saline injection. The power of the placebo effect lies in the mind's ability to influence physical health, demonstrating the complex interplay between psychological and physiological processes.

The placebo effect is widely recognized and studied in clinical trials, where it is used as a control to test the effectiveness of new medications and treatments. By comparing the results of a drug to those of a placebo, researchers can determine the actual efficacy of the treatment. This phenomenon not only highlights the importance of patient belief and expectation in the healing process but also raises intriguing questions about the potential of the mind to influence bodily functions and the perception of pain.

> *"The mind is a powerful tool in healing." — John F. Kennedy*

Fascinating Study on the Placebo Effect

One notable study on the placebo effect was conducted by researchers at Harvard Medical School and published in the Journal of Neuroscience in 2004. The study aimed to understand how the brain responds to placebo treatments by using advanced imaging techniques. Participants with chronic pain were divided into two groups: one received a placebo treatment, and the other received no treatment. The researchers used functional magnetic resonance imaging (fMRI) to observe changes in brain activity.

The study found that participants who received the placebo treatment reported significant pain relief. The fMRI scans revealed that these participants also showed increased activity in brain regions associated with pain relief, such as the anterior cingulate cortex and the prefrontal cortex. These findings suggested that the placebo effect was not merely a subjective experience but involved actual physiological changes in the brain.

The researchers concluded that the placebo effect could activate endogenous pain relief mechanisms in the brain, similar to how actual medications work. This study provided compelling evidence that the mind's belief in treatment can lead to real, measurable changes in brain function and pain perception.

Did you know?

Studies have shown that the placebo effect can be so strong that it sometimes produces side effects, known as the "nocebo effect," where patients experience negative symptoms because they believe they are taking a harmful or unpleasant treatment.

The Study Where Participants Got Drunk

In a fascinating and somewhat unconventional study published in 2006 in the journal Human Factors, researchers set out to compare the dangers of drunk driving with those of using a mobile phone while driving. To do this, they recruited 40 social drinkers and put their driving skills to the test under various conditions using a driving simulator.

The Experiment Setup

The study involved four separate driving sessions for each participant:

- **Handheld Mobile Phone Use**: In the first session, participants drove while using a handheld mobile phone.
- **Hands-Free Mobile Phone Use:** In the second session, they drove while talking on a hands-free mobile phone.
- **Legally Drunk:** In the third session, participants were served free cocktails until they reached the legal blood alcohol limit for driving, then tested on their driving abilities.
- **Sober and Undistracted:** The fourth session tested participants' driving performance when they were completely sober and not using any mobile phone.

The results of the study were eye-opening. Drivers using either type of mobile phone—handheld or hands-free—were involved in significantly more traffic accidents in the driving simulation compared to when they were drunk or sober. In fact, mobile phone users were more than five times as likely to cause an accident as nondistracted drivers.

These findings underscore the severe dangers of mobile phone use while driving, often underappreciated compared to the well-known risks of drunk driving.

The study highlighted that even hands-free devices, often considered safer, did not reduce the risk of accidents. This research has important implications for public safety and has contributed to the ongoing discussion about stricter regulations and public awareness campaigns to discourage mobile phone use while driving.

Did You Know?

Memory Blackouts Alcohol-induced blackouts occur when the blood alcohol concentration (BAC) reaches levels high enough to impair the hippocampus, the brain region responsible for forming new long-term memories. There are two types of blackouts: "en bloc" blackouts, which result in a complete loss of memory for events, and "fragmentary" blackouts, which involve partial memory loss where some fragments can be recalled with cues. Despite the inability to form new memories, individuals can still engage in complex behaviors such as holding conversations, driving, or even continuing to drink because short-term or working memory remains intact. High levels of alcohol inhibit the brain's glutamate system, which is essential for memory encoding, disrupting communication between neurons and preventing the consolidation of short-term memories into long-term storage.

Repeated episodes of alcohol-induced blackouts can lead to more severe cognitive impairments and increase the risk of developing alcohol dependence. Chronic heavy drinking may also cause long-term damage to the hippocampus, affecting memory and learning abilities.

Black Holes

Black holes are one of the most fascinating and enigmatic phenomena in the universe. Formed from the remnants of massive stars that have undergone gravitational collapse, black holes possess gravitational fields so intense that nothing, not even light, can escape their pull.

This immense gravitational force creates a boundary known as the event horizon, beyond which any matter or radiation is inexorably drawn inward, never to return. The core of a black hole, known as the singularity, is a point of infinite density where the laws of physics as we know them cease to apply.

- There are three main types of black holes: stellar-mass black holes, formed from collapsing stars; supermassive black holes, found at the centers of galaxies, including our own Milky Way; and intermediate-mass black holes, which are less common and not as well understood.
- Black holes cannot be observed directly because they emit no light. However, their presence is inferred by the effect of their gravity on nearby stars and gas. For instance, the orbits of stars near the center of our galaxy suggest the presence of a supermassive black hole.
- Named after physicist Stephen Hawking, this theoretical prediction suggests that black holes can emit radiation due to quantum effects near the event horizon. This process could eventually lead to the evaporation of black holes over incredibly long timescales.
- Matter that gets too close to a black hole forms an accretion disk as it spirals inward. The intense gravitational forces heat this material to extremely high temperatures.
- The largest known black holes, supermassive black holes, can have masses equivalent to billions of suns. The supermassive black hole at the center of the Milky Way, known as Sagittarius A*, has a mass about 4 million times that of our Sun.
- If an object gets too close to a black hole, it experiences spaghettification, where the gravitational forces stretch it into a long, thin shape. This dramatic effect is due to the extreme

difference in gravitational pull between the object's closer and farther sides.
- Some theoretical physicists suggest that black holes could be gateways to other parts of the universe or even to other universes, forming what is known as wormholes. While this remains speculative, it adds to the intrigue surrounding black holes.
- Black holes can create ripples in the fabric of spacetime known as gravitational waves, especially when they collide. These waves, first detected by the LIGO observatory in 2015, have opened a new way of observing and understanding the universe.

Did You Know?

The first image of a black hole was captured in 2019 by the Event Horizon Telescope, a network of radio telescopes around the world. This groundbreaking image showed the shadow of the supermassive black hole in the galaxy M87, providing direct visual evidence of these mysterious cosmic objects.

"Black holes are where God divided by zero." — Steven Wright

Debunking Space Myths

Myth 1: The Great Wall of China is Visible from Space

Fact: Contrary to popular belief, the Great Wall of China is not visible to the naked eye from space. While it is long, the wall is not wide enough to be distinguishable from the Earth's orbit. Astronauts in low Earth orbit can see city lights, major highways, and large structures like airports, but not the Great Wall.

Myth 2: Space is Completely Silent

Fact: While it is true that space is a near-vacuum and lacks an atmosphere to carry sound waves like on Earth, it is not entirely silent. Astronauts and spacecraft can still hear sounds through vibrations transmitted through solid objects, such as their spacecraft. Additionally, instruments on satellites can detect radio waves from various cosmic sources.

Myth 3: You Explode in Space Without a Suit

Fact: If exposed to the vacuum of space without a suit, you wouldn't explode, but you would be in serious danger. Your body would swell due to the lack of external pressure, and you would lose consciousness within seconds due to lack of oxygen. However, immediate death would not occur; it would take about 1-2 minutes for irreversible damage to occur due to oxygen deprivation.

Myth 4: The Moon Has a Dark Side

Fact: The term "dark side of the Moon" is a misnomer. Both sides of the Moon experience day and night due to its rotation. The far side of the Moon, which always faces away from Earth, is not perpetually dark; it receives sunlight just as the near side does. The correct term is the "far side of the Moon."

Myth 5: Space Travel Causes Rapid Aging

Fact: Space travel does expose astronauts to increased levels of radiation and the effects of microgravity, which can have various impacts on the body. However, there is no evidence to suggest that it causes rapid aging. In fact, time dilation, a concept from Einstein's theory of relativity, implies that astronauts traveling at high speeds would actually age more slowly relative to people on Earth.

Myth 6: Black Holes Suck Everything In

Fact: Black holes do have strong gravitational pulls, but they do not act like cosmic vacuum cleaners. Objects must come very close to a black hole to be pulled in. If our Sun were replaced by a black hole of the same mass, Earth's orbit would remain unchanged because the gravitational pull at that distance would be the same.

Myth 7: The Sun is Yellow

Fact: The Sun actually emits white light, which is a combination of all visible wavelengths. It appears yellow from Earth due to the scattering of shorter blue wavelengths in our atmosphere. In space, the Sun appears white, not yellow.

Did you know?

Space is not completely devoid of matter. The interstellar medium, the matter that exists in the space between the stars, consists of gas (mostly hydrogen and helium) and dust particles. Although it is extremely sparse, it plays a crucial role in the formation of stars and planets.

Space in Cinema: Reality vs. Fiction

Space has always fascinated filmmakers, inspiring a plethora of movies that attempt to capture the vastness and mystery of the cosmos. While these films often captivate audiences with stunning visuals and thrilling narratives, they frequently take liberties with scientific accuracy to enhance the drama and excitement. This results in a mix of movies that range from highly realistic portrayals to those that significantly deviate from the principles of space science. Here, we explore three movies praised for their realism and three that are notorious for their scientific inaccuracies.

Three Good Movies for Space Realism

1. Interstellar (2014)

Directed by Christopher Nolan, "Interstellar" is renowned for its commitment to scientific accuracy, particularly in its depiction of black holes and wormholes. The film's visual representation of a black hole, based on input from physicist Kip Thorne, has been praised for its fidelity to current astrophysical theories. The narrative explores the complexities of space travel, time dilation, and the survival of humanity in a future where Earth is becoming uninhabitable.

2. The Martian (2015)

Ridley Scott's adaptation of Andy Weir's novel "The Martian" focuses on an astronaut stranded on Mars and his efforts to survive until rescue. The film is celebrated for its attention to detail, accurately portraying the challenges of life on Mars, including the need for sustainable food and water sources and the use of scientific ingenuity to solve life-threatening problems. NASA scientists consulted on the film to ensure its scientific credibility.

3. Apollo 13 (1995)

Directed by Ron Howard, "Apollo 13" is a dramatic retelling of the real-life 1970 Apollo 13 lunar mission, which faced a critical failure en route to the Moon. The film is lauded for its accurate depiction of the mission's technical challenges and the resourcefulness of the astronauts and ground crew. By sticking closely to historical events and real NASA protocols, "Apollo 13" offers a compelling and authentic glimpse into space exploration.

Three Bad Movies for Space Realism

1. Armageddon (1998)

Directed by Michael Bay, "Armageddon" is often cited as one of the least scientifically accurate space movies. The plot revolves around sending oil drillers to an asteroid to prevent it from colliding with Earth. The film is filled with scientific inaccuracies, from the portrayal of space travel to the feasibility of the mission's concept. Critics and scientists alike have pointed out numerous errors, making it more of a space fantasy than a realistic depiction.

2. Gravity (2013)

Although "Gravity," directed by Alfonso Cuarón, is visually stunning and received critical acclaim, it suffers from several scientific inaccuracies. The film's depiction of orbital mechanics, distances between space stations, and the effects of microgravity are often exaggerated or incorrect. Despite these flaws, "Gravity" remains a gripping narrative but should be viewed with an understanding of its scientific liberties.

> *"Movies are not about reality; they are about storytelling. But sometimes, a good story needs to respect the truths of the universe."*

3. Space Cowboys (2000)

Directed by Clint Eastwood, "Space Cowboys" tells the story of aging astronauts sent to repair an old Soviet satellite. The film stretches the bounds of realism with its portrayal of space technology and the physical capabilities of older astronauts undertaking such a demanding mission. While entertaining, it sacrifices scientific accuracy for dramatic effect and nostalgic appeal.

Time Dilation Explained: Real Life and "Interstellar"

Time dilation is a concept from Einstein's theory of relativity that describes how time can pass at different rates for observers depending on their relative velocities and the strength of gravitational fields they experience. There are two types of time dilation: velocity-based (special relativity) and gravitational (general relativity).

According to Einstein's special theory of relativity, time moves slower for objects moving at high speeds compared to those at rest. This effect becomes significant at speeds approaching the speed of light. And according to Einstein's general theory of relativity, time moves slower in stronger gravitational fields. This means that the closer an object is to a massive source of gravity, the slower time will pass for it relative to an object further away from the gravitational source.

Time dilation is also observed in particle accelerators, where particles are accelerated to speeds close to the speed of light. These particles have longer lifetimes than they would at rest, as time slows down for them due to their high velocities.

Astronauts on the International Space Station (ISS) experience a slight time dilation effect because they are traveling at high speeds around the Earth. Over the course of six months, the time difference is minuscule (only milliseconds), but it is measurable.

Plot Context

In "Interstellar," time dilation plays a crucial role in the storyline, particularly in the scenes involving the planet Miller, which orbits close to the massive black hole Gargantua.

Gravitational Time Dilation

Miller's Planet: The planet Miller is so close to Gargantua that the gravitational field is extremely strong. According to general relativity, time passes much more slowly on Miller compared to Earth. In the film, it is stated that one hour on Miller's planet is equivalent to seven years on Earth. This dramatic difference in time passage is a direct consequence of gravitational time dilation.

Visual and Plot Representation

- The film accurately depicts the effects of gravitational time dilation with stunning visual effects. As the crew approaches Miller's planet, the massive black hole Gargantua looms large, demonstrating the intense gravitational forces at play.
- The crew's mission to Miller's planet results in significant time loss. When they return to their spacecraft, they discover

that 23 years have passed for the crew member left on board, even though only a few hours have passed for those who visited the planet. This profound time difference underscores the emotional and scientific implications of time dilation.

Famous inventors and their inventions

1 - Who invented the first practical refrigerator?

    A. Thomas Edison
    B. Carl von Linde
    C. Albert Einstein
    D. James Harrison

2 - Which inventor is credited with creating the first programmable computer?

    A. Charles Babbage
    B. Alan Turing
    C. Konrad Zuse
    D. John Atanasoff

3 - Who developed the process of pasteurization?

    A. Marie Curie
    B. Louis Pasteur
    C. Alexander Fleming
    D. Gregor Mendel

4 - The inventor of the modern alternating current (AC) electricity supply system was:

    A. Nikola Tesla
    B. Thomas Edison
    C. Benjamin Franklin

5 - Who is known for inventing dynamite?

    A. Alfred Nobel
    B. Antoine Lavoisier

    C. Fritz Haber
    D. Robert Oppenheimer

6 - Which inventor created the first successful photographic process, the daguerreotype?

    A. George Eastman
    B. Louis Daguerre
    C. Nicéphore Niépce
    D. Thomas Wedgwood

Answers

1. D) James Harrison
2. C) Konrad Zuse
3. B) Louis Pasteur
4. A) Nikola Tesla
5. A) Alfred Nobel
6. B) Louis Daguerre

Human Genome Project

The Human Genome Project was an international effort to sequence and map all the genes - together known as the genome - of members of our species, Homo sapiens. Completed in 2003, the project aimed to identify all the approximately 20,000-25,000 genes in human DNA, determine the sequences of the 3 billion chemical base pairs that make up human DNA, store this information in databases, improve tools for data analysis, transfer related technologies to the private sector, and address the ethical, legal, and social issues (ELSI) that may arise from the project.

1990: Officially launched in October, the HGP was expected to last 15 years but, due to rapid technological advances, completed all primary goals ahead of schedule in 2003.

2000: A working draft of the genome was announced, creating a buzz worldwide about the potential medical and scientific implications.

2003: The project was declared complete on April 14, 2003, providing a roadmap to more than 99% of the human genome's gene-containing regions with 99.99% accuracy.

The HGP has accelerated our understanding of mutations linked to different types of diseases, leading to new strategies for diagnosis, treatment, and, in some cases, prevention or cure.

With the completion of the HGP, the medical industry began moving towards more personalized approaches to healthcare, where medical treatment can be tailored based on individual genetic profiles. Researchers initially estimated that humans might have over 100,000 genes but found that we have about 20,000-25,000—roughly the same number as a fruit fly.

Type 1 and Type 2 Thinking

Type 1 Thinking: Intuitive and Fast

Type 1 thinking operates automatically and quickly, with little or no effort and no sense of voluntary control. It involves making split-second decisions based on instinct and gut reactions.

Examples: Recognizing faces, understanding simple sentences, and detecting hostility in someone's voice are all tasks managed by Type 1 thinking.

Advantages: It helps us respond quickly in situations where speed is more valuable than accuracy, such as avoiding immediate dangers.

Limitations: While efficient, Type 1 thinking can lead to biases and errors because it relies heavily on associative memory and familiar patterns, often ignoring logical analysis.

Type 2 Thinking: Analytical and Slow

Type 2 thinking allocates attention to effortful mental activities that demand it, including complex computations. The operations are often slower, associated with the subjective experience of agency, choice, and concentration.

Examples: Focusing on a challenging math problem, evaluating the pros and cons of a major decision, and engaging in strategic planning are all driven by Type 2 thinking.

Advantages: This type of thinking allows for deeper and more rational analysis, leading to more accurate judgments and decisions in complex situations.

Limitations: Type 2 thinking requires more energy and time, making it less practical for everyday decisions and often overridden by the faster and more automatic Type 1 thinking in routine situations.

Interaction Between Type 1 and Type 2 Thinking

Balancing Act

In daily life, both types of thinking interact seamlessly. Type 1 offers a first reaction to stimuli, which Type 2 can either endorse by taking no action or override by engaging in deeper analysis and questioning.

Error Checking

Type 2 thinking often serves as a check on Type 1 processes, stepping in when decisions require more accuracy, such as revising initial judgments or solving problems that arise from intuitive responses.

"When faced with a difficult question, we often answer an easier one instead, usually without noticing the substitution."

Neurological Bases: Studies using functional MRI scans have shown that different brain regions are activated during Type 1 and Type 2 thinking. For instance, the amygdala plays a crucial role in fast, emotion-driven responses typical of Type 1 thinking, while the frontal lobes are more active during the deliberate and analytical tasks of Type 2 thinking.

Decision Fatigue: Engaging in prolonged periods of Type 2 thinking can lead to decision fatigue, a state where the brain tires and begins to seek shortcuts. This often results in either decision avoidance or a reversion to Type 1 thinking, which can be less taxing but more prone to errors.

> *"We easily think associativity, we think metaphorically, we think casually, but stats requires thinking about many things at once, which is something that System 1 (fast thinking) is not designed to do."*

Impact of Stress: Stress can significantly impact the balance between Type 1 and Type 2 thinking. Under high stress, the brain is more likely to favor Type 1 thinking for quick decision-making, which can be beneficial in immediate danger but detrimental in complex, non-emergency situations.

> *"Intelligence is not only the ability to reason; it's also the ability to find relevant material in memory and to deploy attention when needed.*

Age and Cognitive Development: The efficiency and preference for Type 1 versus Type 2 thinking can change with age. Children and older adults tend to rely more on Type 1 thinking, while adults in their prime working years may be more adept at engaging in Type 2 thinking.

"A reliable way to make people believe in falsehoods is frequent repetition, because familiarity is not easily distinguished from truth.

Quantum Entanglement

Quantum entanglement is one of the most fascinating and puzzling aspects of quantum physics. Coined by Einstein to express his skepticism about the phenomenon, quantum entanglement defies classical notions of physical laws and has profound implications for the future of technology and our understanding of the universe.

Basic concept

Quantum entanglement occurs when pairs or groups of particles interact in ways such that the quantum state of each particle cannot be described independently of the state of the others, even when the particles are separated by large distances.

- Creation of Entanglement: Occurs when particles, such as photons, interact or collide, causing their properties to become interconnected.
- Linking Properties: After entanglement, properties like spin or polarization of one particle are instantaneously linked to those of its partner.
- Non-local Connection: Measurement of a property on one entangled particle instantly sets the corresponding property of the other, regardless of the distance separating them.

"Spooky action at a distance."

EPR Paradox: Highlighted the strange implications of entanglement and questioned the completeness of quantum mechanics.

Bell's Theorem, proposed by physicist John Bell in 1964, is a profound result in the foundations of quantum mechanics that challenges the classical views of reality. The theorem addresses one of the central puzzles in quantum physics—quantum entanglement—and whether it can be explained within the framework of classical physics

Bell's theorem centers around quantum entanglement, where two or more particles become so linked that the state of one (regardless of distance) instantaneously affects the state of the other. Bell investigated whether the predictions of quantum mechanics could be accounted for by 'local hidden variables.' These are hypothetical, unseen variables that supposedly predetermine the outcomes of quantum measurements, in accordance with classical physics principles that require influences to travel at speeds not exceeding the speed of light (the principle of locality).

Quantum Teleportation: Utilizes entanglement to transmit the state of a particle to another location without physical transfer.

Quantum Cryptography: Uses entanglement to ensure secure communication by detecting eavesdropping attempts.

Superdense Coding: Allows transmission of two classical bits of information using one entangled qubit.

Alain Aspect's Experiments: Provided strong experimental evidence against local hidden variable theories in the 1980s.

Quantum Computing: Relies on entanglement to enable qubits to perform parallel computations, increasing processing power.

Multiple choice Quiz on physics

What is the unit of measure for force in the International System of Units (SI)?

    A. Newton
    B. Joule
    C. Pascal

Which principle explains why a boat floats on water?

    A. Pascal's Principle
    B. Archimedes' Principle
    C. Newton's Third Law

What is the term for the speed at which a projectile must travel to break free of Earth's gravitational pull?

    A. Terminal velocity
    B. Escape velocity
    C. Critical speed

Which of the following is not a type of fundamental force in the universe?

    A. Gravitational
    B. Elastic

C. Electromagnetic

In the context of thermodynamics, what does the zeroth law primarily relate to?

A. Conservation of energy
B. Equilibrium temperature
C. Increase of entropy

Which phenomenon demonstrates the particle nature of light?

A. Diffraction
B. Interference
C. Photoelectric effect

Answers - A) Newton B) Archimedes' Principle B) Escape velocity B) Elastic B) Equilibrium temperature C) Photoelectric effect

"Physics is not about how the world is, it is about what we can say about the world." — Niels Bohr

"The universe is under no obligation to make sense to you." — Neil deGrasse Tyson

"Everything is theoretically impossible, until it is done." — Robert A. Heinlein

"In physics, you don't have to go around making trouble for yourself—nature does it for you." — Frank Wilczek

"There are no secrets about the world of nature. There are secrets about the thoughts and intentions of men." — J. Robert Oppenheimer

True Or False - Physics

True or False: The speed of light in a vacuum is approximately 300,000 kilometers per second.

True or False: In a vacuum, all objects fall at the same rate regardless of their mass.

True or False: The First Law of Thermodynamics states that energy cannot be created or destroyed, only transferred or changed in form.

True or False: In quantum mechanics, Heisenberg's Uncertainty Principle states that you can simultaneously know the exact position and exact momentum of a particle.

True or False: A catalyst increases the rate of a chemical reaction by increasing the activation energy.

True or False: Sound travels faster in water than in air.

True or False: The gravitational force between two objects increases as the distance between them increases.

True or False: In a vacuum, electromagnetic waves travel at the same speed regardless of their frequency.

Answers

True or False: The speed of light in a vacuum is approximately 300,000 kilometers per second.

True

The exact speed of light in a vacuum is 299,792 kilometers per second (about 186,282 miles per second). This is a fundamental constant of nature and is denoted by the symbol - c

True or False: In a vacuum, all objects fall at the same rate regardless of their mass.

True

In a vacuum, there is no air resistance, so all objects fall at the same rate under gravity. This principle was famously demonstrated by Galileo and later confirmed by experiments conducted on the Moon by Apollo astronauts.

True or False: The First Law of Thermodynamics states that energy cannot be created or destroyed, only transferred or changed in form.

True

The First Law of Thermodynamics, also known as the Law of Energy Conservation, states that the total energy of an isolated system is constant; energy can be transformed from one form to another, but cannot be created or destroyed.

True or False: In quantum mechanics, Heisenberg's Uncertainty Principle states that you can simultaneously know the exact position and exact momentum of a particle.

False

Heisenberg's Uncertainty Principle states that it is impossible to simultaneously know both the exact position and exact momentum of a particle.

True or False: A catalyst increases the rate of a chemical reaction by increasing the activation energy.

False

A catalyst increases the rate of a chemical reaction by lowering the activation energy required for the reaction to occur. It provides an alternative reaction pathway with a lower activation energy.

True or False: Sound travels faster in water than in air.

True

Sound travels faster in water than in air because water molecules are more closely packed together than air molecules, allowing sound waves to transfer energy more efficiently. The speed of sound in water is about 1,480 meters per second, compared to approximately 343 meters per second in air.

True or False: The gravitational force between two objects increases as the distance between them increases.

False

The gravitational force between two objects decreases as the distance between them increases. According to Newton's law of universal gravitation, the force is inversely proportional to the square of the distance between the two objects.

True or False: In a vacuum, electromagnetic waves travel at the same speed regardless of their frequency.

True

In a vacuum, all electromagnetic waves travel at the speed of light (approximately 299,792 kilometers per second) regardless of their frequency or wavelength. This speed is a constant and is one of the fundamental properties of electromagnetic radiation.

Cosmic Alchemy

Cosmic alchemy refers to the process by which elements are formed and transformed in the universe through nuclear fusion and other stellar processes. This term beautifully encapsulates the idea that stars act as cosmic forges, where simpler elements like hydrogen and helium are fused into heavier elements, creating the diverse array of elements found throughout the cosmos.

Big Bang Nucleosynthesis

Shortly after the Big Bang, the universe was incredibly hot and dense, allowing the formation of the lightest elements. During the first few minutes, protons and neutrons combined to form the nuclei of hydrogen (including deuterium), helium, and small amounts of lithium and beryllium. These elements constituted the primordial matter from which stars and galaxies eventually formed.

Stellar Nucleosynthesis:

Stars are the primary sites of element formation in the universe. In the cores of stars, nuclear fusion reactions convert hydrogen into helium, releasing energy in the process. As stars evolve, they can fuse heavier elements in their cores

Lifecycle of Stars

- Main Sequence Stars: Fuse hydrogen into helium.
- Red Giants and Supergiants: Fuse helium into carbon and oxygen, and in more massive stars, create even heavier elements up to iron.

Elements up to iron are formed in the cores of stars during their lifetimes.

"We are all made of star-stuff." — Carl Sagan

Supernova Nucleosynthesis

When massive stars exhaust their nuclear fuel, they undergo supernova explosions. These cataclysmic events generate extreme temperatures and pressures, allowing for the creation of elements heavier than iron through rapid neutron capture processes (r-process).

- Supernovae scatter these newly formed elements into space, enriching the interstellar medium with heavy elements.

Neutron Star Mergers

Recent discoveries have shown that the collision and merger of neutron stars can also produce heavy elements through the r-process. These mergers are powerful enough to forge elements like gold, platinum, and other heavy metals.

- Neutron star mergers contribute significantly to the abundance of heavy elements in the universe.

Implications of Cosmic Alchemy:

- The elements produced through cosmic alchemy are the building blocks of planets and life. Elements like carbon, oxygen, nitrogen, and phosphorus are essential for the formation of organic molecules and living organisms.
- The process of cosmic alchemy explains the chemical diversity observed in the universe. From the iron in our blood to the gold in jewelry, these elements were forged in the hearts of stars and distributed through stellar explosions and mergers.
- Studying cosmic alchemy helps astronomers and astrophysicists understand the life cycles of stars, the formation of galaxies, and the evolution of the universe. It provides insights into the processes that have shaped the cosmos over billions of years.

Supernova Nucleosynthesis

When massive stars exhaust their nuclear fuel, they undergo supernova explosions. These cataclysmic events generate extreme temperatures and pressures, allowing for the creation of elements heavier than iron through rapid neutron capture processes (r-process).

Outcome: Supernovae scatter these newly formed elements into space, enriching the interstellar medium with heavy elements.

Neutron Star Mergers

Recent discoveries have shown that the collision and merger of neutron stars can also produce heavy elements through the r-process. These mergers are powerful enough to forge elements like gold, platinum, and other heavy metals.

Outcome: Neutron star mergers contribute significantly to the abundance of heavy elements in the universe.

Formation of Planets and Life

The elements produced through cosmic alchemy are the building blocks of planets and life. Elements like carbon, oxygen, nitrogen, and phosphorus are essential for the formation of organic molecules and living organisms.

Chemical Diversity: The process of cosmic alchemy explains the chemical diversity observed in the universe. From the iron in our blood to the gold in jewelry, these elements were forged in the hearts of stars and distributed through stellar explosions and mergers.

Understanding the Universe: Studying cosmic alchemy helps astronomers and astrophysicists understand the life cycles of stars, the formation of galaxies, and the evolution of the universe. It provides insights into the processes that have shaped the cosmos over billions of years.

Isaac Newton

A towering figure in the history of science, Sir Isaac Newton made groundbreaking contributions that laid the foundation for classical mechanics and profoundly influenced mathematics and physics. Here are some fascinating facts about this extraordinary scientist:

Laws of Motion:

Newton formulated the three laws of motion, which describe the relationship between a body and the forces acting upon it, and the body's motion in response to those forces. These laws are fundamental to our understanding of movement and mechanics.

First Law: An object will remain at rest or in uniform motion unless acted upon by an external force (also known as the law of inertia).

Second Law: The acceleration of an object is directly proportional to the net force acting on it and inversely proportional to its mass ($F = ma$).

Third Law: For every action, there is an equal and opposite reaction.

Universal Gravitation

Newton's law of universal gravitation posits that every particle of matter in the universe attracts every other particle with a force that is directly proportional to the product of their masses and inversely proportional to the square of the distance between their centers. This law explains the force that keeps planets in orbit around the sun and governs the motion of celestial bodies.

> *"To myself, I am only a child playing on the beach, while vast oceans of truth lie undiscovered before me."*

Mathematics: Newton, independently of Gottfried Wilhelm Leibniz, developed the field of calculus, a branch of mathematics that deals with continuous change. His work in calculus provided tools essential for scientific advancements.

Optics: Newton made significant contributions to optics, including the discovery that white light is composed of a spectrum of colors. He demonstrated this by passing light through a prism, showing that it could be split into its constituent colors and recombined into white light.

Reflecting Telescope: In 1668, Newton invented the reflecting telescope, which used a curved mirror instead of lenses to reflect light and form an image. This design reduced chromatic aberration and improved the quality of the images.

Alchemy and Theology: Beyond his scientific endeavors, Newton was deeply interested in alchemy and spent a considerable amount of time studying ancient texts and conducting alchemical experiments. He also wrote extensively on theological topics, exploring the nature of God and the interpretation of biblical prophecies.

Childhood and Education: Born prematurely on December 25, 1642, in Woolsthorpe, England, Newton was not expected to survive. Despite a challenging childhood, he excelled academically and attended Trinity College, Cambridge, where he began his groundbreaking work.

Publication of Principia: Newton's seminal work, "Philosophiæ Naturalis Principia Mathematica" (Mathematical Principles of Natural Philosophy), commonly known as the Principia, was first published in 1687. This work outlined his laws of motion and universal gravitation and is considered one of the most important works in the history of science.

Royal Society Presidency: Newton served as the president of the Royal Society from 1703 until his death in 1727. During his tenure, he influenced the direction of scientific research and ensured the society's prominence in the scientific community.

- Newton had a unique and often reclusive personality. He was known for being intensely focused on his work, to the point of neglecting food and sleep. He also had a famously contentious relationship with fellow scientist Robert Hooke.
- Newton's reflecting telescope was not just an improvement over existing designs; it was the first practical model that used mirrors instead of lenses to avoid chromatic aberration, a common issue in telescopes at the time.

- Legend has it that Newton's dog, Diamond, caused a fire in his laboratory that destroyed 20 years of research. While the story's accuracy is debated.
- The famous story of Newton being inspired to formulate his theory of gravity by watching an apple fall is partially true. He did observe an apple fall, which led him to ponder the forces at work, but the idea of gravity was developed over many years of contemplation and experimentation.

Carl Linnaeus

Known as the "father of modern taxonomy," Carl Linnaeus revolutionized the classification of living organisms. His work laid the foundation for the binomial nomenclature system we use today, drastically improving the organization and communication of biological information.

Contributions to Science

- Binomial Nomenclature: Linnaeus developed the system of binomial nomenclature, assigning each species a two-part Latin name consisting of the genus and species. This system provided a standardized way to name and classify organisms, reducing confusion in scientific communication.
- Systema Naturae: In 1735, Linnaeus published the first edition of "Systema Naturae," a comprehensive work that classified thousands of plant and animal species. The book went through many editions, each expanding and refining his classification system.
- Classification Hierarchy: Linnaeus introduced a hierarchical structure for classifying organisms, including ranks such as kingdom, class, order, family, genus, and species. This hierarchical system is still in use, with some modifications, in modern taxonomy.

- Human Classification: Linnaeus was the first to include humans in his classification of animals, placing them in the genus Homo and the species sapiens. This was a groundbreaking and controversial inclusion at the time.

"God created, Linnaeus organized." — Carl Linnaeus

Legacy in Botany

Linnaeus's work had a profound impact on botany. He described over 7,000 plant species in his lifetime and his classification system is still used by botanists today. Many plants bear names he assigned.

Travel and Exploration

Linnaeus conducted extensive fieldwork, traveling across Sweden and other parts of Europe to collect specimens and observe organisms in their natural habitats. His expeditions contributed significantly to his understanding and classification of biodiversity.

Student Influence

Linnaeus's students, known as "apostles," traveled around the world collecting specimens and spreading his classification system. Their efforts helped establish Linnaeus's methods as the standard in biological sciences.

Linnaean Garden

Linnaeus maintained a botanical garden in Uppsala, Sweden, where he cultivated and studied a wide variety of plants. The garden became an important center for botanical research and education.

Cultural Impact

Linnaeus's work extended beyond science into culture and education. He was a prolific writer and communicator, making his ideas accessible to both scientists and the general public. His classification system was not only a scientific tool but also a means of understanding the natural world.

Noble Title

In recognition of his scientific contributions, Linnaeus was ennobled by the Swedish king in 1761, receiving the name Carl von Linné. This honor reflected his esteemed position in society and his impact on science.

Charles Darwin

A pioneering naturalist and biologist, Charles Darwin fundamentally changed our understanding of life on Earth with his theory of evolution by natural selection. His work laid the groundwork for modern evolutionary biology and has had a profound impact on science, philosophy, and society.

Theory of Evolution: Darwin's theory of evolution by natural selection posits that species evolve over time through a process where individuals with advantageous traits are more likely to survive and reproduce, passing those traits to future generations.

On the Origin of Species: In 1859, Darwin published "On the Origin of Species by Means of Natural Selection," which presented extensive evidence for evolution and introduced the concept of natural selection. This book is considered one of the most important scientific works ever published.

Galápagos Islands: During his voyage on the HMS Beagle, Darwin observed distinct variations in species on the Galápagos Islands. These observations, particularly of finches with different beak shapes, were crucial in developing his theory of natural selection.

Common Descent: Darwin proposed the idea of common descent, which suggests that all species share a common ancestor. This concept is a cornerstone of evolutionary biology and helps explain the relationships between different organisms.

"It is not the strongest of the species that survive, nor the most intelligent, but the one most responsive to change." — Charles Darwin

- Darwin's five-year journey aboard the HMS Beagle (1831-1836) was instrumental in shaping his scientific ideas. He collected a vast array of specimens and made detailed observations of geology, flora, and fauna around the world.
- Although Darwin did not know about genes or DNA, his theory laid the groundwork for the field of genetics. The later discovery of Mendelian inheritance provided the genetic mechanisms that explained how traits are passed from one generation to the next.
- Darwin's ideas sparked significant debate and controversy, particularly among religious groups. The theory of evolution challenged traditional views on the creation of life, leading to ongoing discussions about science and religion.
- Alfred Russel Wallace, a contemporary of Darwin, independently developed a theory of natural selection. Wallace's work prompted Darwin to publish his own findings, and the two scientists presented their theories jointly to the Linnean Society of London in 1858.
- Darwin's work influenced a wide range of fields beyond biology, including psychology, anthropology, and sociology. The concept of evolution has been applied to understand human behavior, social structures, and cultural development.
- Darwin suffered from various health problems throughout his life, including chronic illness that often confined him to his home. Despite these challenges, he continued to work and correspond with other scientists, making significant contributions to science until his death.

The Galápagos Islands

An archipelago located about 1,000 kilometers (620 miles) off the coast of Ecuador, the Galápagos Islands are renowned for their unique wildlife and their role in the development of Charles Darwin's theory of evolution.

Unique Wildlife: The Galápagos Islands are home to numerous species found nowhere else on Earth, including the Galápagos tortoise, marine iguana, and Galápagos penguin. These unique species have adapted to the islands' diverse environments.

Diverse Ecosystems: The archipelago consists of 13 major islands and several smaller islands, each with its own distinct ecosystem. The islands' varying climates and landscapes support a wide range of flora and fauna.

Darwin's Inspiration: Charles Darwin visited the Galápagos Islands in 1835 during his voyage on the HMS Beagle. His observations of the islands' finches, which had different beak shapes adapted to their specific diets, were crucial in forming his theory of natural selection.

Volcanic Origins: The Galápagos Islands are volcanic in origin and are located on the Nazca tectonic plate. The islands are still geologically active, with several volcanic eruptions occurring in recent history.

World Heritage Site: The Galápagos Islands were declared a UNESCO World Heritage Site in 1978 due to their outstanding natural beauty and ecological significance. The designation helps protect the islands' unique biodiversity.

Marine Reserve: Surrounding the islands is the Galápagos Marine Reserve, one of the largest marine protected areas in the world. The reserve is home to diverse marine life, including sharks, rays, and sea turtles.

Research and Conservation: The Charles Darwin Research Station, established in 1964 on Santa Cruz Island, conducts scientific research and conservation efforts to protect the islands' unique ecosystems and species.

Climate Variability: The Galápagos Islands experience climate variability due to ocean currents, particularly the cold Humboldt Current and the warm El Niño Southern Oscillation. These currents affect the islands' weather patterns and marine ecosystems.

Tourism Impact: While tourism is a significant source of revenue for the islands, it also poses challenges for conservation. Efforts are made to manage tourism sustainably to minimize its impact on the delicate ecosystems.

> *"The natural history of these islands is eminently curious, and well deserves attention." — Charles Darwin*

The Galápagos Islands remain a living laboratory for the study of evolution and conservation, continuing to inspire scientists and nature enthusiasts from around the world.

Galileo Galilei

Often referred to as the "father of modern observational astronomy," Galileo Galilei made pioneering contributions to physics, astronomy, and the scientific method. His work laid the groundwork for future scientific discoveries and fundamentally changed our understanding of the universe.

Telescope Improvements: Galileo significantly improved the design of the telescope, allowing him to make detailed observations of celestial bodies. His telescopes had magnifications up to 30 times, far surpassing the capabilities of earlier models.

Astronomical Discoveries: Using his improved telescope, Galileo made several groundbreaking discoveries:

- **Jupiter's Moons:** In 1610, he discovered the four largest moons of Jupiter—Io, Europa, Ganymede, and Callisto—now known as the Galilean moons. This was the

first observation of moons orbiting another planet, challenging the geocentric model of the universe.
- **Phases of Venus:** Galileo observed that Venus exhibited phases similar to the Moon, providing evidence for the heliocentric model proposed by Copernicus.
- **Sunspots:** He observed dark spots on the Sun's surface, known as sunspots, which indicated that the Sun was not a perfect, immutable sphere as previously thought.
- **Milky Way:** Galileo discovered that the Milky Way was composed of countless stars, revealing the vastness of the universe.

Law of Inertia: Galileo's experiments with inclined planes and pendulums led to the formulation of the law of inertia, which states that an object in motion will remain in motion unless acted upon by an external force. This principle laid the foundation for Newton's first law of motion.

"You cannot teach a man anything; you can only help him find it within himself." — Galileo Galilei

Conflict with the Church: Galileo's support for the heliocentric model brought him into conflict with the Roman Catholic Church. In 1633, he was tried by the Inquisition and found "vehemently suspect of heresy." He was forced to recant his views and spent the rest of his life under house arrest.

"E pur si muove": Legend has it that after recanting his heliocentric views, Galileo muttered, "E pur si muove" ("And yet it moves"), asserting the truth of the Earth's motion despite the Church's opposition. While this story is likely apocryphal, it symbolizes Galileo's commitment to scientific truth.

Galileo's Thermometer: Galileo invented an early version of the thermometer, known as Galileo's air thermometer, which used the expansion and contraction of air to measure temperature changes.

Galilean Cannonball Experiment: According to popular accounts, Galileo conducted experiments by dropping spheres of different masses from the Leaning Tower of Pisa to demonstrate that their time of descent was independent of their mass. While the exact details of these experiments are debated, they highlight his innovative approach to studying motion.

House Arrest Contributions: Even under house arrest, Galileo continued his scientific work. He wrote "Two New Sciences," which summarized his findings on kinematics and strength of materials, and it is considered one of his most important works.

The Wonders of the World

The Wonders of the World are remarkable constructions and natural formations that have fascinated people for centuries. They are typically divided into different categories, including the Ancient Wonders, Medieval Wonders, Modern Wonders, and Natural Wonders.

Ancient Wonders of the World

Great Pyramid of Giza (Egypt): The only surviving wonder of the original seven, the Great Pyramid was built as a tomb for the Egyptian Pharaoh Khufu. It is an architectural marvel due to its precise construction and immense size.

Hanging Gardens of Babylon (Iraq): Described as an extraordinary series of tiered gardens containing a wide variety of trees, shrubs, and vines. Their existence and location are subject to historical debate, but they were said to have been built in the ancient city-state of Babylon.

Statue of Zeus at Olympia (Greece): A giant seated figure of Zeus, the king of the Greek gods, made by the sculptor Phidias around 435 BC. It was housed in the Temple of Zeus and was renowned for its grandeur and beauty.

Temple of Artemis at Ephesus (Turkey): A large temple dedicated to the goddess Artemis, completed around 550 BC. It was rebuilt after being destroyed in 356 BC and was famous for its impressive size and ornate decorations.

Mausoleum at Halicarnassus (Turkey): A tomb built for Mausolus, a satrap of the Persian Empire, and his wife, Artemisia, around 350 BC.

It was celebrated for its architectural beauty and the artistry of its sculptures.

Colossus of Rhodes (Greece): A massive bronze statue of the sun god Helios, erected on the island of Rhodes around 280 BC. It stood approximately 33 meters high and was considered one of the tallest statues of the ancient world.

Lighthouse of Alexandria (Egypt): Also known as the Pharos of Alexandria, it was built on the small island of Pharos around 280 BC to guide sailors safely into the busy harbor of Alexandria. It was one of the tallest man-made structures in the world for many centuries.

Medieval Wonders of the World:

Stonehenge (England): A prehistoric monument consisting of a ring of standing stones, each around 13 feet high. The site dates back to around 3000 BC to 2000 BC and remains a mystery in terms of its exact purpose.

Colosseum (Italy): An ancient Roman amphitheater in the center of Rome, built between AD 70-80. It was used for gladiatorial contests and public spectacles such as mock sea battles, animal hunts, and executions.

Catacombs of Kom el Shoqafa (Egypt): A historical archaeological site in Alexandria, known for its blend of Egyptian, Greek, and Roman cultural styles. The catacombs were used as a burial site from the 2nd to the 4th century AD.

Great Wall of China (China): A series of fortifications made of various materials, built along the northern borders of China to protect against invasions. Construction began in the 7th century BC and continued until the 17th century AD.

Porcelain Tower of Nanjing (China): A pagoda constructed in the 15th century during the Ming Dynasty. It was made of white porcelain bricks that reflected sunlight, making it a dazzling sight.

Hagia Sophia (Turkey): Originally a Greek Orthodox Christian basilica, later an Ottoman imperial mosque, and now a museum in Istanbul. It was built in AD 537 and is renowned for its massive dome and stunning architecture.

Leaning Tower of Pisa (Italy): A freestanding bell tower of the cathedral of the Italian city of Pisa. Its unintended tilt, which began during construction in the 12th century, makes it a unique and iconic structure.

New 7 Wonders of the World

Great Wall of China (China): Renowned for its immense scale and historical significance, the wall stretches over 13,000 miles.

Petra (Jordan): An ancient city famous for its rock-cut architecture and water conduit system. It was the capital of the Nabataean Kingdom and is also known as the "Rose City" due to the color of the stone.

Christ the Redeemer (Brazil): A massive statue of Jesus Christ in Rio de Janeiro, standing 98 feet tall atop the Corcovado Mountain. It is a symbol of Christianity and a cultural icon of Brazil.

Machu Picchu (Peru): An Incan citadel set high in the Andes Mountains, built in the 15th century. It is renowned for its sophisticated dry-stone construction and panoramic views.

Chichen Itza (Mexico): A large pre-Columbian archaeological site built by the Maya civilization. The site includes the famous pyramid known as El Castillo.

Roman Colosseum (Italy): An iconic symbol of Imperial Rome, the Colosseum is the largest ancient amphitheater ever built and is still partially standing today.

Taj Mahal (India): A white marble mausoleum built in the mid-17th century by Mughal Emperor Shah Jahan in memory of his beloved wife Mumtaz Mahal. It is celebrated for its stunning architecture and intricate craftsmanship.

Natural Wonders of the World:

Grand Canyon (USA): A colossal canyon carved by the Colorado River, known for its breathtaking size and its intricate and colorful landscape.

Great Barrier Reef (Australia): The world's largest coral reef system, composed of over 2,900 individual reefs and 900 islands. It is home to a diverse range of marine life.

Harbor of Rio de Janeiro (Brazil): Known for its stunning natural setting, including Sugarloaf Mountain, Corcovado Peak, and the famous beaches of Copacabana and Ipanema.

Mount Everest (Nepal/Tibet): The highest mountain in the world, standing at 8,848 meters (29,029 feet). It attracts climbers from all over the globe.

Aurora Borealis (Various locations): Also known as the Northern Lights, this natural light display is predominantly seen in high-latitude regions around the Arctic and Antarctic.

Parícutin Volcano (Mexico): A cinder cone volcano that emerged suddenly in a farmer's cornfield in 1943. It grew rapidly and erupted for nine years.

Victoria Falls (Zambia/Zimbabwe): One of the largest and most famous waterfalls in the world, known locally as "The Smoke that Thunders."

"The Pyramids themselves, doting with age, have forgotten the names of their founders." — Thomas Fuller

"This Colossus bestrides the narrow world like a giant." — William Shakespeare

"Not a piece of architecture, as other buildings are, but the proud passions of an emperor's love wrought in living stones." — Sir Edwin Arnold

"Machu Picchu is a place of mysteries, whispers, and echoes, where ancient stones hold the secrets of a lost civilization."

Inventions Quiz

1. Who invented the telephone?

   A. Alexander Graham Bell
   B. Thomas Edison
   C. Nikola Tesla
   D. Guglielmo Marconi

2. What year was the first successful airplane flight by the Wright brothers?

   A. 1903
   B. 1899
   C. 1912
   D. 1908

3. Who is credited with inventing the light bulb?

   A. Nikola Tesla
   B. Thomas Edison
   C. James Watt
   D. George Westinghouse

4. Which inventor is known for the phonograph?

   A. Alexander Graham Bell
   B. Thomas Edison
   C. Heinrich Hertz
   D. Guglielmo Marconi

5. Who invented the World Wide Web?

    A. Bill Gates

    B. Steve Jobs

    C. Tim Berners-Lee

    D. Larry Page

6. What year was the first iPhone released?

    A. 2005

    B. 2007

    C. 2010

    D. 2012

7. Who invented the first practical telephone?

    A. Thomas Edison

    B. Elisha Gray

    C. Alexander Graham Bell

    D. Nikola Tesla

8. Which invention is attributed to Eli Whitney?

    A. Steam engine

    B. Cotton gin

    C. Telegraph

    D. Sewing machine

9. Who invented the polio vaccine?

    A. Louis Pasteur

    B. Edward Jenner

    C. Alexander Fleming

    D. Jonas Salk

10. Who invented the first practical electric motor?

    A. Nikola Tesla

    B. Thomas Edison

    C. Michael Faraday

    D. D) James Watt

11. Who is credited with inventing the first automobile powered by an internal combustion engine?

    A. Henry Ford
    B. Karl Benz
    C. Gottlieb Daimler
    D. Rudolf Diesel

12. Who developed the first successful printing press?

    A. Johannes Gutenberg
    B. William Caxton
    C. Aldus Manutius
    D. Benjamin Franklin

13. What did James Watt invent that significantly impacted the Industrial Revolution?

    A. Light bulb
    B. Telephone
    C. Steam engine
    D. Cotton gin

14. Who is known for inventing the theory of relativity?

    A. Isaac Newton
    B. Albert Einstein
    C. Galileo Galilei
    D. Niels Bohr

15. Who invented the first practical mechanical computer, known as the Analytical Engine?

    A. Charles Babbage
    B. Alan Turing
    C. John von Neumann
    D. Blaise Pascal

16. Who is credited with inventing the safety elevator?

    A. Alexander Graham Bell
    B. Thomas Edison

C. Elisha Otis

D. George Stephenson

17. Which inventor is known for developing the process of pasteurization?

A. Alexander Fleming

B. Louis Pasteur

C. Robert Koch

D. Joseph Lister

18. Who invented the first successful revolver?

A. Samuel Colt

B. John Browning

C. Richard Gatling

D. Hiram Maxim

19. Who is known for inventing the modern zipper?

A. Elias Howe

B. Whitcomb L. Judson

C. Gideon Sundback

D. Thomas Hancock

20. Who developed the theory of electromagnetism?

A. James Clerk Maxwell

B. Michael Faraday

C. Heinrich Hertz

D. Nikola Tesla

Answers

1. Alexander Graham Bell

2. 1903

3. Thomas Edison

4. Thomas Edison

5. Tim Berners-Lee

6. 2007

7. Alexander Graham Bell

8. Cotton gin

9. Jonas Salk

10. Michael Faraday

11. Karl Benz

12. Johannes Gutenberg

13. Steam engine

14. Albert Einstein

15. Charles Babbage

16. Elisha Otis

17. Louis Pasteur

18. Samuel Colt

19. Gideon Sundback

20. James Clerk Maxwell

Karl Benz and the First Automobile

Karl Benz is credited with inventing the first automobile powered by an internal combustion engine, a groundbreaking achievement that revolutionized transportation. In 1885-1886, he designed and built the Benz Patent-Motorwagen, which featured a single-cylinder, four-stroke engine. This engine, mounted horizontally at the rear of the vehicle, had a displacement of 954 cubic centimeters and produced 0.75 horsepower at 400 revolutions per minute (RPM). The innovative design included a high-voltage electrical system, spark plug ignition, and a water-cooled internal combustion engine. Benz's Motorwagen also had an advanced carburetor for fuel intake, a manual crank starter, and a battery ignition system, which were remarkable engineering feats for the time.

Bertha Benz, Karl's wife, played a crucial role in demonstrating the practicality of the Motorwagen. In 1888, she undertook the first long-distance car trip from Mannheim to Pforzheim, a journey of about 66 miles. This trip was not only a testament to the reliability and functionality of the vehicle but also highlighted several key scientific and engineering principles.

Explanation of Magnetic Storms Caused by Solar Flares and Coronal Mass Ejections

Magnetic storms, or geomagnetic storms, are disturbances in Earth's magnetosphere caused by solar activity. These storms are primarily triggered by two types of solar phenomena: solar flares and coronal mass ejections (CMEs). Solar flares are intense bursts of radiation emanating from the Sun's surface when magnetic energy is suddenly released. These flares can produce a wide range of electromagnetic radiation, from radio waves to X-rays and gamma rays.

Coronal mass ejections, on the other hand, are massive bursts of solar wind and magnetic fields rising above the solar corona or being released into space. When a CME collides with Earth's magnetosphere, it can cause significant disruptions by compressing the magnetic field and injecting energetic particles into the magnetosphere. These interactions result in enhanced currents in the magnetosphere and ionosphere, leading to geomagnetic storms.

The impact of these storms can be profound, affecting satellite operations, communication systems, navigation systems, and even power grids. The last significant geomagnetic storm alert was issued in 2005, when Earth experienced an exceptionally high dose of solar radiation. This event was notable because it delivered the highest dose of radiation measured in five decades. The sudden influx of solar particles and energy can lead to auroras in polar regions, but it can also cause technological disruptions, highlighting the importance of monitoring and understanding solar activity and its effects on our planet.

Space Radiation: Types, Sources, and Effects

Space radiation is a significant concern for both astronauts and electronic equipment in space. The two primary types of space radiation are galactic cosmic rays (GCRs) and solar energetic particles (SEPs). GCRs originate from outside our solar system and are composed mainly of high-energy protons and heavy ions. These particles travel at nearly the speed of light and can penetrate deep into shielding materials. SEPs, on the other hand, are high-energy particles ejected by the Sun during solar flares and coronal mass ejections. SEPs consist primarily of protons, electrons, and alpha particles. Both GCRs and SEPs pose substantial risks due to their high energy and penetrating power.

The primary sources of space radiation are galactic cosmic rays, solar energetic particles, and trapped radiation belts. GCRs originate from supernova explosions and other high-energy events outside our solar system. These cosmic events accelerate particles to near-light speeds, creating highly energetic radiation. SEPs are generated by solar flares and coronal mass ejections from the Sun. During these events, the Sun releases bursts of energetic particles that can reach Earth and other parts of the solar system. Additionally, the Earth's magnetosphere traps charged particles from the solar wind, creating the Van Allen radiation belts, which also contribute to the radiation environment in space.

Space radiation poses significant health risks to astronauts.

Prolonged exposure to high levels of radiation can lead to acute radiation sickness, increased cancer risk, and damage to the central nervous system. GCRs, with their high-energy particles, can penetrate the human body and damage DNA and cells, leading to mutations and increased cancer risk. SEPs, although generally lower in energy compared to GCRs, can deliver large doses of radiation over short periods, particularly during intense solar storms. Mitigating these risks requires effective shielding, careful mission planning to avoid peak solar activity, and monitoring of radiation levels during space missions.

Space radiation also adversely affects electronic equipment and spacecraft systems. High-energy particles can cause single-event effects (SEEs) such as bit flips in memory devices, latch-ups in electronic circuits, and even complete system failures. Cumulative radiation exposure can lead to degradation of materials and electronic components, reducing the lifespan and reliability of spacecraft. Engineers design spacecraft with radiation-hardened components and employ shielding techniques to protect sensitive electronics. Additionally, real-time monitoring of space weather and radiation levels helps manage the operational risks associated with space radiation.

- Spacecraft use materials like polyethylene for shielding because it's rich in hydrogen, which is effective at blocking high-energy protons and other particles.
- Space agencies, such as NASA, set strict radiation exposure limits for astronauts to minimize health risks, using a combination of pre-mission planning, real-time monitoring, and post-mission assessments.

Crazy Things Elon Musk Has Done

Launching a Tesla Roadster into Space

In February 2018, Elon Musk's company SpaceX launched the Falcon Heavy rocket, carrying a cherry-red Tesla Roadster as its payload. The car, which belonged to Musk himself, was sent into orbit around the Sun. The Roadster, complete with a dummy driver named "Starman," was intended as a whimsical test payload and is now traveling through space, showcasing Musk's flair for dramatic and unconventional marketing.

Selling Flamethrowers

In 2018, Musk's company The Boring Company started selling flamethrowers as a limited-edition promotional item. Officially named "Not-A-Flamethrower" to circumvent legal restrictions, 20,000 units were sold out quickly, raising $10 million. The stunt not only generated significant media attention but also highlighted Musk's ability to blend business with humor and novelty.

Building a Mini-Sub for a Thai Cave Rescue

During the 2018 Thai cave rescue mission, where 12 boys and their soccer coach were trapped in a flooded cave, Musk quickly developed a mini-submarine designed to assist in the rescue. Although the device wasn't ultimately used, Musk personally traveled to Thailand and engaged with the rescue efforts, demonstrating his rapid-response engineering capabilities and desire to solve real-world problems.

Smoking Marijuana on Joe Rogan's Podcast:

In September 2018, Musk made headlines by smoking marijuana during a live interview on "The Joe Rogan Experience" podcast. The incident sparked controversy and led to a temporary dip in Tesla's stock price. It also raised questions about Musk's behavior and the impact of his public actions on his companies, but it also underscored his unconventional and unpredictable persona.

Creating the "Hyperloop" Concept

In 2013, Musk unveiled the concept for the Hyperloop, a high-speed transportation system that would move pods through low-pressure tubes at speeds of up to 760 mph. Although Musk didn't initially plan to build the Hyperloop himself, his proposal has inspired multiple companies and research groups to pursue the technology. The Hyperloop idea exemplifies Musk's visionary thinking and willingness to challenge traditional transportation methods.

Creating Neuralink to Connect Brains with Computers

In 2016, Elon Musk co-founded Neuralink, a neurotechnology company focused on developing implantable brain–machine interfaces (BMIs). The goal of Neuralink is to merge the human brain with artificial intelligence, enhancing human cognitive abilities and potentially addressing neurological conditions. In 2020, Musk demonstrated a working prototype implanted in a pig, showcasing the device's ability to monitor brain activity. This ambitious project highlights Musk's futuristic vision and his interest in pushing the boundaries of human potential through advanced technology.

The Science Behind Neuralink

Neuralink is a neurotechnology company founded by Elon Musk with the ambitious goal of developing implantable brain-machine interfaces (BMIs). These devices aim to facilitate direct communication between the human brain and computers, potentially revolutionizing the treatment of neurological disorders and enhancing human capabilities.

Neuralink's device consists of ultra-thin, flexible electrode threads that are implanted into the brain. These threads are connected to a small, implantable chip, called the Link, which processes and transmits neural signals. The device is designed to be minimally invasive, using a specially developed robotic surgical system to implant the electrodes with precision and avoid damage to brain tissue.

Key Components of Neuralink:

Electrode Threads:

These are ultra-thin, flexible wires equipped with numerous electrodes that can detect electrical signals from neurons. The threads are designed to minimize damage and inflammation, making long-term implantation feasible.

The Link Chip:

This chip processes the electrical signals picked up by the electrodes. It can wirelessly transmit data to external devices, allowing for real-time monitoring and interaction.

Robotic Surgeon:

The robotic system developed by Neuralink is crucial for the precise insertion of the electrode threads into the brain. It can place the threads with micron-level accuracy, which is essential to avoid blood vessels and ensure optimal positioning.

The Science and Applications:

Neural Signal Interpretation:

The primary function of Neuralink's technology is to interpret the brain's electrical signals. By understanding these signals, the device can potentially translate thoughts into actions, such as moving a cursor on a screen, controlling prosthetic limbs, or interacting with smart devices.

Treatment of Neurological Disorders:

- Restoring Movement: For individuals with spinal cord injuries or paralysis, Neuralink aims to bypass damaged neural pathways, allowing the brain to communicate directly with muscles or external devices.
- Treating Neurological Conditions: Conditions such as Parkinson's disease, epilepsy, and depression could benefit from Neuralink's ability to monitor and modulate brain activity. By precisely targeting affected areas, the device could provide more effective treatments with fewer side effects.

Enhancing Human Capabilities:

- Cognitive Enhancement: In the long term, Neuralink envisions enhancing cognitive functions. This could include memory augmentation, faster information processing, and even direct brain-to-brain communication.
- Merging with AI: Musk has often spoken about the potential for Neuralink to serve as a bridge between humans and artificial intelligence. By integrating with AI, humans could potentially access vast amounts of information quickly, perform complex calculations instantaneously, and interact with AI systems more intuitively.

Predictions for the Future:

Short-Term (Next 5 Years):

- Clinical Trials: Neuralink is expected to progress through clinical trials to ensure the safety and efficacy of its technology. Initial trials will likely focus on restoring movement and treating neurological disorders.
- Regulatory Approval: Gaining approval from regulatory bodies like the FDA will be a critical step towards making Neuralink's devices available for therapeutic use.

Mid-Term (5-10 Years):

- Widespread Medical Use: Assuming successful trials and regulatory approval, Neuralink devices could become more widely used in medical settings, significantly improving the quality of life for individuals with severe neurological impairments.
- Technological Integration: Integration with other technologies, such as prosthetics, smart home devices, and communication systems, could become more sophisticated, providing seamless control and interaction.

Long-Term (Beyond 10 Years):

- Cognitive Enhancement: As the technology matures, applications for cognitive enhancement may become feasible. This could lead to significant advancements in human capabilities, potentially revolutionizing education, work, and personal communication.
- Human-AI Symbiosis: The ultimate goal of creating a symbiotic relationship between humans and AI could redefine what it means to be human, blending biological and digital intelligence in unprecedented ways.

Gene Therapy for Genetic Disorders

Gene Therapy: An innovative and rapidly advancing field in medicine, gene therapy involves the modification of an individual's genes to treat or cure genetic disorders. By targeting the root cause of these conditions at the molecular level, gene therapy offers the potential for long-term and possibly permanent solutions.

What is Gene Therapy?

Gene therapy involves altering the genes inside a person's cells to treat or prevent disease. This can be done by replacing a faulty gene with a healthy copy, inactivating a malfunctioning gene, or introducing a new or modified gene into the body.

Techniques:

- Gene Replacement Therapy: Inserting a normal gene to replace a mutated gene.
- Gene Silencing: Using techniques like RNA interference (RNAi) to "silence" or inactivate a faulty gene.
- Gene Editing: Utilizing tools like CRISPR-Cas9 to directly correct or modify the DNA sequence of a gene.

How Does It Work?

Vectors:

- Viral Vectors: Modified viruses are commonly used to deliver therapeutic genes into cells because of their ability to efficiently enter cells and integrate genetic material.
- Non-viral Vectors: Methods such as liposomes, nanoparticles, and direct DNA injection are also employed to transfer genes without using viruses.

Process:

- Gene Identification: Identifying the defective or missing gene responsible for the disorder.
- Vector Design: Designing a vector to carry the healthy gene into the patient's cells.
- Delivery: Administering the vector to the patient, typically via injection or infusion.
- Cell Integration: The vector delivers the gene into the patient's cells, where it integrates into the DNA and begins to produce the functional protein.

Applications and Success Stories

Applications:

Gene therapy is being explored for a wide range of genetic disorders, including:

- Cystic Fibrosis: A condition caused by mutations in the CFTR gene, leading to severe respiratory and digestive problems.
- Hemophilia: A disorder where the blood does not clot properly due to the lack of specific clotting factors.

- Duchenne Muscular Dystrophy: A severe muscle-wasting disease caused by mutations in the dystrophin gene.
- Sickle Cell Disease: A blood disorder caused by a mutation in the HBB gene, leading to abnormal red blood cells.

Success Stories:

- Luxturna: The first FDA-approved gene therapy for an inherited disease, Luxturna treats Leber's congenital amaurosis, a rare form of inherited blindness.
- Zolgensma: Approved for the treatment of spinal muscular atrophy (SMA), Zolgensma delivers a functional copy of the SMN1 gene to infants and young children, significantly improving their prognosis.
- Sickle Cell Disease: Ongoing clinical trials have shown promising results, with some patients experiencing a dramatic reduction in disease symptoms after gene therapy.

Challenges and Future Directions

Challenges:

- Safety: Ensuring that gene therapy does not cause unintended effects, such as immune reactions or insertional mutagenesis (disrupting other important genes).
- Delivery: Developing effective and targeted delivery systems that can reach the desired cells and tissues without degradation.
- Cost: Gene therapies can be extremely expensive, making them inaccessible for many patients without significant financial assistance or healthcare coverage.

Future Directions:

- CRISPR-Cas9 and Beyond: Advances in gene editing technologies promise more precise and efficient ways to correct genetic defects.
- Broader Applications: Expanding the use of gene therapy to treat more common diseases, such as cancer, heart disease, and diabetes.
- Personalized Medicine: Tailoring gene therapy approaches to the genetic makeup of individual patients for more effective treatments.

"The potential for gene therapy is immense; it holds the promise of treating the root cause of genetic disorders at their source, transforming the landscape of medicine and offering hope to millions."

Did you know?

The first approved gene therapy trial occurred in 1990, where researchers successfully treated a young girl with severe combined immunodeficiency (SCID) by introducing a healthy copy of the ADA gene into her immune cells. This groundbreaking trial paved the way for the development of modern gene therapies.

Deep-Sea Biodiversity

The deep sea, often defined as ocean depths below 200 meters, is one of the most mysterious and least explored environments on Earth. This dark, cold, and high-pressure realm covers about 65% of the Earth's surface and is home to a remarkable diversity of life. Despite the extreme conditions, deep-sea ecosystems are incredibly rich and complex, hosting a variety of organisms that have adapted in extraordinary ways to survive in their unique habitats. From hydrothermal vent communities to the abyssal plains, the deep sea is teeming with life forms that display fascinating adaptations, such as bioluminescence, chemosynthesis, and gigantism.

Life in the deep sea has evolved numerous adaptations to cope with the harsh environment. Bioluminescence, the production of light by living organisms, is widespread among deep-sea creatures. This adaptation is used for various purposes, including attracting prey, deterring predators, and communicating. Another key adaptation is chemosynthesis, a process where certain bacteria convert chemicals from hydrothermal vents into energy, forming the basis of the food web in these dark regions. Additionally, many deep-sea species exhibit gigantism, where organisms grow to larger sizes compared to their shallow-water relatives, possibly due to slower metabolism rates and reduced predation pressures.

The deep sea plays a critical role in the global ecosystem. It acts as a major carbon sink, helping to regulate the Earth's climate by sequestering carbon dioxide from the atmosphere. Deep-sea organisms contribute to nutrient cycling and the decomposition of organic matter, maintaining the health of marine ecosystems. Furthermore, the deep sea is a reservoir of genetic diversity, with many species yet to be discovered, offering potential benefits for biotechnology and medicine. Understanding and preserving deep-sea biodiversity is essential, especially in the face of increasing threats such as deep-sea mining, pollution, and climate change, which could have far-reaching impacts on the entire planet.

- Bioluminescence: Over 90% of deep-sea organisms are capable of bioluminescence, producing their own light through chemical reactions.
- Hydrothermal Vents: Discovered in 1977, hydrothermal vents are rich sources of minerals and support unique ecosystems where life thrives without sunlight.
- Pressure Adaptations: Deep-sea creatures have flexible cell membranes and specialized proteins to function under extreme pressures exceeding 1,000 times atmospheric pressure.

- Gigantism: Examples of deep-sea gigantism include the giant squid, which can grow up to 13 meters, and the Japanese spider crab, with a leg span of up to 3.8 meters.
- Deep-Sea Exploration: Only about 20% of the ocean floor has been mapped with high-resolution imaging, leaving much of the deep sea unexplored and unknown.
- Methane Seeps: These are areas where methane gas escapes from the seabed, supporting diverse communities of organisms that rely on chemosynthetic bacteria for energy.

Memory Formation and Retention

Memory formation is a complex process that involves encoding, storing, and retrieving information. This process begins with encoding, where sensory information is transformed into a format that can be stored in the brain. This involves attention and perception, where information from our senses is processed and integrated. Once encoded, the information is stored in various regions of the brain, primarily in the hippocampus, which plays a crucial role in consolidating short-term memories into long-term memories. Over time, these memories are distributed to other parts of the brain for long-term storage, particularly in the cerebral cortex.

Retention of memory is influenced by several factors, including emotional significance, repetition, and the context in which the information was learned. Emotionally charged events tend to be remembered more vividly due to the involvement of the amygdala, which enhances the encoding of these memories. Repetition and practice, such as through study or rehearsal, strengthen neural connections, making it easier to retrieve the information later. Additionally, the context and environment in which the memory was formed can serve as cues for recall. This phenomenon, known as context-dependent memory, suggests that individuals are more likely to remember information when they are in the same environment where they initially learned it.

Forgetting is a natural part of memory retention and occurs due to several mechanisms. One common reason is decay, where memories weaken over time if they are not accessed or rehearsed. Another mechanism is interference, where similar pieces of information compete with each other, leading to confusion or loss of detail. There are two types of interference: proactive interference, where old memories interfere with the formation of new ones, and retroactive interference, where new information makes it harder to recall older memories. Additionally, retrieval failure, which can happen when the brain is unable to access stored information due to a lack of retrieval cues, is another reason for forgetting. Understanding these mechanisms can help in developing strategies to improve memory retention and recall.

- Sleep and Memory: Sleep plays a critical role in memory consolidation. During sleep, especially during the REM stage, the brain processes and solidifies memories.
- Neuroplasticity: The brain's ability to reorganize itself by forming new neural connections throughout life is known as neuroplasticity, which is fundamental to learning and memory.
- Mnemonics: Mnemonic devices are techniques that aid memory retention by associating new information with familiar patterns, such as acronyms or rhymes.
- Exercise and Memory: Regular physical exercise has been shown to enhance memory and cognitive function by increasing blood flow to the brain and promoting the growth of new neurons.
- Diet and Memory: Certain foods rich in antioxidants, omega-3 fatty acids, and vitamins can support brain health and improve memory function.
- Stress and Memory: Chronic stress can negatively impact memory formation and retention by damaging the hippocampus and impairing cognitive function.

The Science of Sleep

Which hormone is primarily responsible for regulating sleep-wake cycles?

    A. Insulin
    B. Melatonin
    C. Cortisol

What is the average recommended amount of sleep for an adult per night?

    A. 5-6 hours
    B. 7-9 hours
    C. 10-12 hours

Which stage of sleep is most associated with vivid dreams?

    A. REM sleep
    B. NREM Stage 2
    C. NREM Stage 3

What term describes the natural, internal process that regulates the sleep-wake cycle and repeats roughly every 24 hours?

    A. Homeostasis
    B. Circadian rhythm
    C. Metabolism

Which of the following is NOT a characteristic of REM sleep?

    A. Rapid eye movement
    B. Muscle atonia (temporary paralysis)
    C. Deep, restorative sleep

What percentage of sleep time is typically spent in REM sleep for a healthy adult?

    A. 10-15%
    B. 20-25%
    C. 30-35%

Which sleep disorder is characterized by difficulty falling or staying asleep?

    A. Narcolepsy
    B. Insomnia
    C. Sleep apnea

During which stage of sleep does the body perform the most repair and growth?

    A. NREM Stage 1
    B. REM sleep
    C. NREM Stage 3

What is the term for the phenomenon where sleep-deprived individuals experience brief, involuntary episodes of sleep?

    A. Microsleep
    B. Sleepwalking
    C. Sleep paralysis

Which neurotransmitter is known to promote wakefulness and alertness?

    A. Serotonin
    B. Dopamine
    C. Norepinephrine

Answers: The Science of Sleep

Which hormone is primarily responsible for regulating sleep-wake cycles?

    B) Melatonin

What is the average recommended amount of sleep for an adult per night?

    B) 7-9 hours

Which stage of sleep is most associated with vivid dreams?

A) REM sleep

What term describes the natural, internal process that regulates the sleep-wake cycle and repeats roughly every 24 hours?

B) Circadian rhythm

Which of the following is NOT a characteristic of REM sleep?

C) Deep, restorative sleep

What percentage of sleep time is typically spent in REM sleep for a healthy adult?

B) 20-25%

Which sleep disorder is characterized by difficulty falling or staying asleep?

B) Insomnia

During which stage of sleep does the body perform the most repair and growth?

C) NREM Stage 3

What is the term for the phenomenon where sleep-deprived individuals experience brief, involuntary episodes of sleep?

A) Microsleep

Which neurotransmitter is known to promote wakefulness and alertness?

C) Norepinephrine

Albert Einstein

Albert Einstein's Journey from Germany to the United States

Albert Einstein was born on March 14, 1879, in Ulm, in the Kingdom of Württemberg in the German Empire. He grew up in Munich, where he later attended the Luitpold Gymnasium. In 1896, Einstein enrolled at the Swiss Federal Polytechnic in Zurich, where he studied physics and mathematics. After graduating in 1900, he worked at the Swiss Patent Office in Bern while developing his groundbreaking theories. In 1905, Einstein published four seminal papers in the "Annalen der Physik," including the special theory of relativity and the famous equation $E=mc^2$, which earned him recognition as a leading scientist.

Einstein's work gained international acclaim, and in 1914, he accepted a position at the Prussian Academy of Sciences in Berlin and became a professor at the University of Berlin. During this period, he developed the general theory of relativity, published in 1915, which revolutionized our understanding of gravity. Despite his scientific success, Einstein's life in Germany became increasingly difficult due to the political climate. A pacifist and a Jew, Einstein faced growing anti-Semitism and political unrest in the wake of World War I. He became an outspoken advocate for civil rights and criticized the rising tide of nationalism and militarism in Germany.

With the rise of Adolf Hitler and the Nazi Party in 1933, the situation in Germany deteriorated rapidly for Jews and political dissidents. Einstein, who was visiting the United States when Hitler came to power, decided not to return to Germany. The Nazis had seized his property, and he was publicly denounced as an enemy of the state. In response, Einstein accepted a position at the Institute for Advanced Study in Princeton, New Jersey. In 1933, he formally emigrated to the United States, where he continued his scientific work and became an American citizen in 1940.

In the United States, Einstein used his prominence to advocate for various causes, including Zionism, civil rights, and nuclear disarmament. He continued his research in theoretical physics, contributing to the development of quantum mechanics and the unified field theory. During World War II, Einstein famously signed a letter to President Franklin D. Roosevelt, warning of the potential for Nazi Germany to develop atomic weapons and urging the development of similar weapons in the U.S., which led to the Manhattan Project. Einstein spent the remainder of his life in Princeton, where he passed away on April 18, 1955. His journey from Germany to the United States is a testament to his resilience and enduring impact on both science and society.

"If you can't explain it simply, you don't understand it well enough."

Albert Einstein's Theories

1. Special Theory of Relativity (1905):

- Proposed that the speed of light is constant in all inertial frames of reference.
- Introduced the concept that time and space are interwoven into a single continuum known as spacetime.
- Resulted in the famous equation $E=mc^2$, showing the relationship between mass and energy.

2. General Theory of Relativity (1915):

- Described gravity as a curvature of spacetime caused by mass and energy.
- Predicted phenomena such as the bending of light around massive objects (gravitational lensing).
- Provided the theoretical foundation for the later discovery of black holes and the expanding universe.

3. Photoelectric Effect (1905):

- Demonstrated that light can be understood as both a wave and a particle (photon).
- Showed that electrons are emitted from materials when they absorb photons with sufficient energy.
- This work provided crucial evidence for quantum theory and earned Einstein the Nobel Prize in Physics in 1921.

4. Brownian Motion (1905):

- Explained the erratic movement of particles suspended in a fluid as a result of collisions with atoms or molecules.
- Provided empirical evidence for the existence of atoms and molecules, supporting atomic theory.
- This theory helped to establish statistical mechanics and the kinetic theory of gases.

5. Cosmological Constant (1917):

- Introduced a term in the equations of general relativity to allow for a static universe.
- Later abandoned the idea after the discovery of the universe's expansion by Edwin Hubble.
- The cosmological constant was revived in modern cosmology to explain dark energy and the accelerated expansion of the universe.

6. Unified Field Theory:

- An effort to unify the fundamental forces of nature, specifically electromagnetism and gravity.
- Sought to create a single theoretical framework that could describe all physical phenomena.
- Although Einstein did not complete this theory, it laid the groundwork for future research in theoretical physics, such as string theory.

"Imagination is more important than knowledge. For knowledge is limited, whereas imagination embraces the entire world, stimulating progress, giving birth to evolution."

Did You Know? Crazy Facts About Albert Einstein

Einstein's Brain Travels

After his death in 1955, Albert Einstein's brain was removed without his family's permission for scientific study. It was divided into 240 pieces and preserved, with pieces traveling around for research for many years.

Missing Socks

Einstein famously hated wearing socks and often went without them. He found socks to be a nuisance and would go out of his way to avoid wearing them, even with formal attire.

Late Speech Development

Einstein didn't start speaking until he was about four years old, and his parents were worried he might have a learning disability. His delayed speech led to what is now known as "Einstein Syndrome," a term used to describe late-talking children who are exceptionally gifted.

Failed University Entrance Exam

At the age of 16, Einstein failed the entrance exam to the Swiss Federal Polytechnic in Zurich. He excelled in math and science but did poorly in other subjects like language and history. He eventually passed the following year after additional preparation.

Offer for Presidency

In 1952, after the death of Israel's first president, Chaim Weizmann, Einstein was offered the presidency of Israel. He respectfully declined, stating that he lacked the necessary experience and people skills for the position.

Illegitimate Daughter

Einstein and his first wife, Mileva Maric, had an illegitimate daughter named Lieserl, born in 1902. Very little is known about her fate, and she is believed to have either died young or been adopted.

Complicated Love Life

Einstein married his first cousin, Elsa Löwenthal, after divorcing his first wife, Mileva Maric. His marriage to Elsa was also complex, as it was rumored that Einstein had numerous affairs, which he admitted to in letters.

Travel Document Controversy

In 1933, when Einstein was traveling to the United States, his travel documents and luggage were confiscated by the Nazis. They even raided his home, and the German government branded him an enemy of the state.

Inventor of a Refrigerator:

Einstein co-invented a refrigerator with his former student, Leo Szilard, in 1926. Known as the Einstein-Szilard refrigerator, it had no moving parts and operated at constant pressure, using ammonia, butane, and water. Although it was never widely commercialized, it demonstrated Einstein's innovative thinking beyond theoretical physics.

"I have no special talent. I am only passionately curious."

What is Nanotechnology?

Nanotechnology is the science, engineering, and application of materials and devices with structures on the nanometer scale, typically between 1 and 100 nanometers (one billionth of a meter). At this incredibly small scale, materials often exhibit unique physical, chemical, and biological properties that differ significantly from their larger-scale counterparts. These properties can include enhanced strength, lighter weight, increased chemical reactivity, or improved electrical conductivity, making nanotechnology a versatile and transformative field. Researchers and engineers utilize these properties to develop new technologies and materials that can revolutionize a wide range of industries, from medicine and electronics to energy and environmental science.

The applications of nanotechnology are vast and varied, encompassing everything from targeted drug delivery systems in medicine to the creation of more efficient solar panels in renewable energy. In the medical field, for instance, nanoparticles can be designed to deliver drugs directly to cancer cells, minimizing damage to healthy tissue and improving treatment efficacy. In electronics, nanotechnology has enabled the development of smaller, faster, and more efficient components, such as transistors and memory chips. Additionally, nanomaterials are being explored for environmental applications, including water purification and pollution control. As research continues to advance, the potential uses and benefits of nanotechnology are expected to grow, offering innovative solutions to some of the world's most pressing challenges.

- Many modern sunscreens contain nanoparticles of zinc oxide or titanium dioxide. These nanoparticles provide effective UV protection while being transparent on the skin, unlike traditional sunscreen ingredients.

- Scientists are developing nanorobots, often called nanobots, that can perform tasks inside the human body, such as delivering drugs to specific cells or repairing damaged tissues at a molecular level.
- Graphene, a single layer of carbon atoms arranged in a two-dimensional lattice, is one of the strongest materials known, despite being only one atom thick. It is approximately 200 times stronger than steel.
- Nanotechnology has led to the creation of self-cleaning materials that mimic the lotus leaf's surface, which repels water and dirt. These materials are used in various applications, including self-cleaning windows and stain-resistant fabrics.
- Nanotechnology is used to create paints that are not only more durable and resistant to chipping but also have self-cleaning properties and can change color with temperature or light exposure.
- Carbon nanotubes are incredibly strong and lightweight materials with extraordinary electrical conductivity. They are used in a variety of applications, including strengthening materials, enhancing battery performance, and even in medical implants.
- Advances in nanotechnology have led to the development of high-resolution imaging techniques, such as atomic force microscopy (AFM) and scanning tunneling microscopy (STM), allowing scientists to visualize and manipulate individual atoms and molecules.

"There's plenty of room at the bottom."
— *Richard Feynman*

The Evolution of 21st Century Private Space Companies

In the 21st century, private space companies have revolutionized the aerospace industry by driving innovation, reducing costs, and expanding the possibilities of space travel. Here's a look at some of the leading private space companies and their contributions:

SpaceX (Elon Musk):

Founded in 2002 by Elon Musk, SpaceX has become a pioneer in commercial space travel and exploration. The company's mission is to reduce space transportation costs and enable the colonization of Mars.

- Reusable Rockets: SpaceX's Falcon 9 and Falcon Heavy rockets are designed for reusability, significantly lowering the cost of space missions.
- Crew Dragon: Developed under NASA's Commercial Crew Program, the Crew Dragon spacecraft has successfully transported astronauts to and from the International Space Station (ISS).
- Starship: Currently in development, Starship is a fully reusable spacecraft intended for missions to Mars and beyond.

SpaceX aims to make space travel more accessible and sustainable, with long-term goals including the colonization of Mars and the development of commercial space travel for private citizens.

2. Blue Origin (Jeff Bezos):

Founded in 2000 by Jeff Bezos, Blue Origin focuses on building reusable rocket technology to make space travel more affordable and sustainable.

- New Shepard: A suborbital rocket designed for space tourism, capable of carrying passengers to the edge of space and back.
- New Glenn: An orbital launch vehicle currently under development, designed to carry heavy payloads and reusable for multiple flights.
- Lunar Missions: Blue Origin is developing the Blue Moon lander to support NASA's Artemis program and potential lunar colonization.

Blue Origin aims to create a future where millions of people live and work in space, developing the infrastructure needed for long-term space habitation.

3. Virgin Galactic (Richard Branson):

Founded by Richard Branson in 2004, Virgin Galactic focuses on commercial space tourism, aiming to make space accessible to non-professional astronauts.

- SpaceShipTwo: A suborbital spaceplane designed to carry passengers to the edge of space, offering a few minutes of weightlessness.
- Commercial Flights: Virgin Galactic has successfully conducted crewed test flights and is preparing to launch regular commercial space tourism flights.

Virgin Galactic aims to democratize space travel by providing suborbital spaceflights for private individuals, researchers, and educational purposes.

4. Rocket Lab (Peter Beck):

Founded in 2006 by Peter Beck, Rocket Lab focuses on providing cost-effective and frequent access to space for small satellites and research missions.

- Electron Rocket: A small, reusable launch vehicle designed to carry small satellites into orbit.
- Mission Successes: Rocket Lab has successfully launched numerous missions, deploying satellites for various commercial, research, and government customers.

Rocket Lab aims to revolutionize the small satellite launch market, making space more accessible for research, commercial, and government purposes.

5. Sierra Nevada Corporation (Eren and Fatih Ozmen):

Sierra Nevada Corporation (SNC) is known for developing advanced aerospace systems and technologies. Their space division focuses on creating innovative solutions for space exploration.

- Dream Chaser: A reusable spaceplane designed for cargo resupply missions to the ISS, capable of landing on conventional runways.
- Lunar Gateway: SNC is contributing to NASA's Lunar Gateway project, which aims to establish a sustainable human presence on the Moon.

SNC aims to develop cutting-edge technologies to support space exploration, including reusable spacecraft and lunar infrastructure.

The Future of Space Travel

The 21st century has seen remarkable advancements in space travel, driven by private space companies such as SpaceX, Blue Origin, Virgin Galactic, Rocket Lab, and Sierra Nevada Corporation. These companies are pushing the boundaries of what is possible, making space more accessible by reducing costs, increasing the frequency of launches, and developing reusable technologies. The future of space travel holds exciting possibilities, including commercial space tourism, lunar bases, and even human missions to Mars.

2040: The Dawn of Space Tourism and Lunar Exploration

By 2040, space tourism is expected to become a reality for those who can afford it. Companies like Virgin Galactic and Blue Origin are pioneering suborbital flights, offering private citizens the chance to experience a few minutes of weightlessness and breathtaking views of Earth from the edge of space. Elon Musk's SpaceX plans to have sent the first crewed missions to Mars by this time, aiming to establish the foundation for a human colony. Additionally, lunar bases are likely to be under construction, serving as both scientific research stations and stepping stones for further space exploration. Jeff Bezos has expressed his vision of millions of people living and working in space, with Blue Origin playing a crucial role in making this a reality through its lunar missions and reusable rockets.

2050: Permanent Lunar Bases and Expanding Space Tourism

By 2050, the establishment of permanent lunar bases is expected to be well underway. These bases will serve as hubs for scientific research, resource extraction, and even tourism. The moon's resources, such as water ice, could be used to produce fuel and support life, making extended stays on the lunar surface feasible. Space tourism will have expanded beyond suborbital flights to include longer stays in orbit, possibly in space hotels. Companies like SpaceX and Blue Origin aim to have regular missions to Mars, with the goal of building a self-sustaining colony. Musk envisions a city on Mars with a population of up to one million people, while Bezos sees space colonies in orbit as a solution to overpopulation and resource depletion on Earth.

2060: Human Settlements on Mars and Advanced Space Habitats

Looking towards 2060, human settlements on Mars may become a reality, thanks to the efforts of SpaceX and other private space companies. These settlements will be designed to be self-sustaining, with infrastructure for living, working, and producing food and energy. Advanced space habitats, both on Mars and in orbit, will incorporate cutting-edge technologies for life support, radiation protection, and sustainable living. Space tourism will have become more mainstream, with space hotels offering extended stays and unique experiences such as spacewalks and lunar excursions. The collaboration between private companies and international space agencies will be crucial in overcoming the challenges of long-term human presence in space.

2070: Interplanetary Travel and Space-Based Economies

By 2070, interplanetary travel could become more routine, with regular flights between Earth, the moon, and Mars. The development of space-based economies will accelerate, with industries such as mining, manufacturing, and energy production operating in space. The establishment of spaceports and logistics hubs will facilitate the movement of people and goods between celestial bodies. Space tourism will continue to grow, with destinations including the moon, Mars, and even asteroids. The visions of Elon Musk and Jeff Bezos—humanity as a multiplanetary species and millions of people living and working in space—will begin to materialize. Their pioneering efforts will have laid the groundwork for a new era of space exploration, where human presence extends beyond Earth, creating opportunities for scientific discovery, economic growth, and the expansion of civilization into the cosmos.

Crazy Facts About the Future of Space Travel

Space Hotels:

- Orion Span's Aurora Station: Expected to launch in the late 2020s, Aurora Station will offer guests a 12-day stay in space for around $9.5 million. Guests will experience zero gravity, stunning views of Earth, and the thrill of living in space.
- Bigelow Aerospace's B330: This inflatable habitat module is designed to expand in space to provide more living space. Bigelow aims to lease these modules for scientific research and tourism.

Space Mining:

- Asteroid Mining Ventures: Companies like Planetary Resources and Deep Space Industries are developing technologies to mine asteroids for precious metals like platinum and essential resources like water, which can be converted into rocket fuel.
- Economic Potential: The value of minerals and water in a single asteroid could be worth trillions of dollars, revolutionizing both the space economy and industries on Earth.

Mars Colonization:

- SpaceX's Mars Plan: Elon Musk's vision includes sending the first crewed mission to Mars in the next decade, with the long-term goal of creating a thriving, self-sustaining city of up to one million people by 2050. This city would have homes, schools, shops, and everything needed to support human life.
- Terraforming Mars: Musk has even discussed the idea of terraforming Mars to make it more Earth-like, potentially using nuclear explosions at the poles to release CO_2 and create a thicker atmosphere.

Lunar Tourism:

- Lunar Base Camp: NASA and international partners are working on the Artemis program to establish a sustainable human presence on the moon by the 2030s. This base could serve as a destination for lunar tourists.
- Tourist Activities: Future tourists could participate in guided moonwalks, visit historic Apollo landing sites, and stay in habitats with amenities similar to Earth.

Reusable Rockets:

- Cost Reduction: SpaceX's Starship aims to lower the cost of space travel from millions of dollars per kilogram to just a few hundred dollars. This dramatic reduction could make space travel accessible to more people and industries.
- Multiple Missions: Starship is designed to be reused many times, similar to how commercial airplanes operate, further reducing costs and increasing the frequency of space missions.

Spaceports:

A. Global Spaceports: SpaceX, Blue Origin, and other companies are planning to build spaceports at strategic locations around the world. These spaceports will serve as launch and landing sites for space missions, facilitating easier and more frequent access to space.
B. Infrastructure Development: Spaceports will include facilities for rocket maintenance, passenger training, and payload preparation, supporting the growth of the space travel industry.

3D-Printed Habitats:

- Local Resources: By using local materials such as lunar regolith or Martian soil, 3D printing can create sturdy structures without the need to transport building supplies from Earth. This approach significantly reduces costs and logistics challenges.
- Habitat Designs: Companies like ICON and AI SpaceFactory are developing 3D printing technologies for constructing habitats that can withstand the harsh conditions of space environments.

Space Farming:

- Controlled Environment Agriculture: NASA and private companies are researching how to grow food in controlled environments with limited resources. Techniques include hydroponics, aeroponics, and using LED lighting to optimize plant growth.
- Sustainable Diets: Future space missions and colonies will rely on space farming to provide fresh produce and ensure a balanced diet for astronauts and settlers.

Zero-Gravity Sports:

- New Sports Leagues: The unique conditions of zero gravity open up possibilities for entirely new sports that take advantage of weightlessness, such as zero-gravity basketball or floating obstacle courses.
- Space Arenas: Specialized arenas could be built on space stations or in lunar and Martian colonies, providing entertainment and recreational activities for both residents and tourists.

Space Junk Cleanup:

- Active Debris Removal: Companies and organizations are developing technologies to capture and remove space debris. Methods include using robotic arms, nets, harpoons, and lasers to target and deorbit debris.
- Ensuring Safety: Cleaning up space debris is critical for the safety of current and future missions, preventing collisions that could damage satellites, spacecraft, and space stations.

Top 5 Science-Focused Countries in the World

1. United States:

- Innovation and Research: The United States leads in scientific research and innovation, with numerous prestigious institutions like MIT, Harvard, and Caltech. It is a global leader in fields such as biotechnology, aerospace, computer science, and medical research.
- Funding and Resources: The U.S. allocates significant funding to research and development through agencies like NASA, NIH, NSF, and DARPA, supporting groundbreaking discoveries and technological advancements.
- Nobel Prizes: The U.S. has produced a high number of Nobel laureates in sciences, reflecting its strong emphasis on cutting-edge research and scientific excellence.

2. China:

- Rapid Growth: China has rapidly increased its investment in science and technology, emerging as a global leader in various fields, including quantum computing, telecommunications, and renewable energy.
- Research Output: China is now one of the top countries in terms of research publications and patents, with substantial contributions to scientific literature.
- Government Support: The Chinese government strongly supports scientific development through initiatives like the Made in China 2025 plan and substantial funding for R&D.

3. Germany:

- Engineering and Technology: Germany is renowned for its engineering prowess and technological innovation, particularly in automotive, industrial machinery, and chemical industries.

- Research Institutions: Institutions like the Max Planck Society, Fraunhofer Society, and prestigious universities like the University of Munich drive significant scientific research and development.
- Collaborative Research: Germany excels in collaborative research, both within Europe and internationally, fostering advancements across multiple scientific disciplines.

4. Japan:

- Advanced Technologies: Japan is a global leader in robotics, electronics, and materials science, with companies like Sony, Toyota, and Panasonic at the forefront of innovation.
- Research and Development: Japan invests heavily in R&D, with a focus on practical applications and technological advancements that improve quality of life.
- Nobel Laureates: Japanese scientists have made significant contributions to fields such as physics, chemistry, and medicine, earning numerous Nobel Prizes.

5. South Korea:

- Technological Innovation: South Korea is known for its advancements in information technology, semiconductor manufacturing, and biotechnology, driven by companies like Samsung and LG.
- Government Initiatives: The South Korean government prioritizes science and technology, investing in research infrastructure and fostering a culture of innovation.
- Educational Excellence: South Korea's strong emphasis on education, particularly in STEM fields, contributes to its scientific achievements and technological advancements.

Small Countries with Interesting Science and Technology Projects

Switzerland:

- CERN (European Organization for Nuclear Research): Although an international collaboration, CERN is based in Switzerland and is home to the Large Hadron Collider (LHC), the world's largest and most powerful particle accelerator. This facility is at the forefront of particle physics research, including the discovery of the Higgs boson.
- EPFL (École Polytechnique Fédérale de Lausanne): This leading engineering school is involved in cutting-edge research in areas such as robotics, biomedical engineering, and renewable energy.

Singapore:

- Biopolis: A research and development hub dedicated to biomedical sciences. It hosts a range of institutes and companies working on genomics, biotechnology, and pharmaceuticals.
- Smart Nation Initiative: Singapore is investing heavily in becoming a "Smart Nation" through the use of advanced technologies in urban planning, healthcare, and transportation. Projects include autonomous vehicles, smart homes, and nationwide sensor networks.

Israel:

- Silicon Wadi: Known as the Silicon Valley of the Middle East, Israel is a global leader in technology and startups. It has significant projects in cybersecurity, artificial intelligence, and agricultural technology.

- Iron Dome: A mobile air defense system developed to intercept and destroy short-range rockets and artillery shells. It is a testament to Israel's advanced defense technology capabilities.

Finland:

- OuluHealth: An innovation hub focusing on digital health and smart health solutions. It combines cutting-edge research with practical applications in healthcare.
- 5G Research: Finland is at the forefront of developing and implementing 5G technology, with extensive research and pilot projects aimed at creating robust and fast mobile networks.

Estonia:

- e-Estonia: Estonia is a pioneer in digital governance, offering an extensive array of e-services to its citizens, including e-residency, online voting, and digital healthcare records.
- Cybersecurity: Estonia has a strong focus on cybersecurity, with the NATO Cooperative Cyber Defence Centre of Excellence located in Tallinn, driving research and innovation in cyber defense.

Denmark:

- Green Energy Projects: Denmark is a leader in renewable energy, particularly wind energy. The country has ambitious plans to become carbon neutral by 2050, with significant investments in offshore wind farms and sustainable energy solutions.
- BioInnovation Institute: This initiative supports life sciences startups and research projects, focusing on areas like bioengineering, medical technology, and biotechnology.

Time Travel in Theory and Popular Culture

Time travel has long been a subject of fascination and speculation within the scientific community. Here are some key theories:

General Relativity:

Proposed by Albert Einstein, the theory of General Relativity revolutionized our understanding of time and space. According to this theory, time and space are intertwined into a four-dimensional continuum called spacetime. Massive objects, such as stars and planets, warp the fabric of spacetime, creating what we perceive as gravity.

- Closed Timelike Curves (CTCs): One of the intriguing implications of General Relativity is the possibility of closed timelike curves. These are paths in spacetime that return to the same point in space and time, theoretically allowing an object to travel back in time. This concept is most commonly associated with rotating black holes, known as Kerr black holes. If a Kerr black hole's rotation is fast enough, it might create a region of spacetime where CTCs exist, potentially enabling time travel.

Wormholes:

Wormholes are hypothetical tunnels through spacetime that could connect distant points in space and time. They are also known as Einstein-Rosen bridges, named after the scientists who proposed them.

- Einstein-Rosen Bridges: These bridges are solutions to the equations of General Relativity that describe a tunnel-like structure connecting two separate points in spacetime. In theory, a wormhole could allow for instantaneous travel between distant regions of space and possibly time. However, the stability of these structures is a major question. Wormholes might require exotic matter with negative energy density to remain open, something that has yet to be observed or created.

Quantum Mechanics:

Quantum mechanics, the branch of physics dealing with the behavior of particles at the smallest scales, offers some intriguing possibilities for time travel.

- Many-Worlds Interpretation: One interpretation of quantum mechanics suggests that every possible outcome of a quantum event actually occurs, resulting in a vast number of parallel universes. According to this interpretation, time travel could involve moving between these parallel universes rather than traveling within a single timeline. This could theoretically bypass many of the paradoxes associated with traditional time travel concepts.

Cosmic Strings:

Cosmic strings are hypothetical one-dimensional defects in the fabric of spacetime, proposed in some models of the early universe. These strings could be remnants of the universe's formation.

- Time Machine Effect: If two cosmic strings were to pass by each other at near the speed of light, they could create significant distortions in spacetime. The interaction of their immense gravitational fields might produce conditions where closed timelike curves could form, potentially allowing for time travel. This concept remains highly speculative and theoretical, as cosmic strings have not been observed.

"The UFO phenomenon exists. The phenomenon is real. I have always wanted to know more about it."

— Dr. J. Allen Hynek

Time Travel in Media and Literature

Time travel has been a popular theme in media and literature, captivating audiences with its imaginative possibilities. Here are some notable examples:

- **H.G. Wells' "The Time Machine" (1895):** Often considered the first major work of fiction to explore time travel, this novel introduced the concept of using a machine to travel through time.
- **"Doctor Who" (1963-present):** This long-running British TV series features the Doctor, a Time Lord who travels through time and space in the TARDIS, a time machine that looks like a British police box. The series explores numerous historical and futuristic settings.
- **"Back to the Future" Trilogy (1985-1990):** These iconic films follow the adventures of Marty McFly and Doc Brown as they travel to the past and future using a DeLorean car converted into a time machine.
- **"Primer" (2004):** This indie film delves into the complexities and paradoxes of time travel with a focus on realism and scientific accuracy. It presents a more grounded and technically plausible depiction of time travel's challenges.
- **"Outlander" Series by Diana Gabaldon (1991-present):** These novels, and their TV adaptation, combine historical fiction with time travel. The protagonist, Claire Randall, is transported from 1945 to 1743 Scotland, where she navigates the dangers and complexities of the past.
- **"Interstellar" (2014):** Directed by Christopher Nolan, this film incorporates advanced scientific concepts, including time dilation and wormholes, to explore the possibilities of time travel in the context of space exploration.

"It is entirely possible that behind the perception of our senses, worlds are hidden of which we are unaware."

— **Albert Einstein**

UFO Phenomena and Government Disclosure

Historical UFO Sightings

The fascination with unidentified flying objects (UFOs) dates back centuries, but modern interest surged in the 20th century with several high-profile sightings.

"The distinction between the past, present, and future is only a stubbornly persistent illusion."

— Albert Einstein

The Roswell Incident (1947)

Perhaps the most famous UFO event, the Roswell Incident, occurred in July 1947 near Roswell, New Mexico.

- Initial Reports: A rancher named William "Mac" Brazel discovered debris scattered across his property. Initial reports described the debris as a "flying disc" or "flying saucer." This led to widespread media coverage and speculation about extraterrestrial visitation.
- Military Explanation: Shortly after the discovery, the U.S. military issued a statement claiming the debris was from a weather balloon. However, many found this explanation unsatisfactory, especially after the military retracted the statement, attributing the debris to a classified project known as

"Project Mogul," which involved high-altitude balloons designed to detect Soviet nuclear tests.

- Conspiracy Theories: The Roswell Incident has since fueled numerous conspiracy theories. Some believe that the government recovered a crashed alien spacecraft and bodies of extraterrestrial beings, which were subsequently hidden in secret facilities like Area 51. Despite numerous investigations and official explanations, the incident remains a focal point for UFO enthusiasts and conspiracy theorists.

The Kenneth Arnold Sighting (1947):

On June 24, 1947, private pilot Kenneth Arnold reported seeing nine unusual flying objects while flying near Mount Rainier, Washington.

- The Sighting: Arnold described the objects as crescent-shaped, traveling at incredible speeds estimated to be over 1,200 miles per hour. He compared their movement to "a saucer if you skip it across the water," which led to the media coining the term "flying saucer."

- Impact on Public Perception: Arnold's sighting is considered the first widely publicized UFO sighting in the United States and marked the beginning of the modern era of UFO sightings. It sparked a wave of similar reports across the country and increased public interest in unidentified aerial phenomena.

- Skeptical Explanations: Some skeptics suggest that Arnold may have misidentified known aircraft or atmospheric phenomena. However, no definitive explanation has been provided, leaving the sighting open to interpretation and speculation.

The Phoenix Lights (1997):

On the night of March 13, 1997, thousands of people witnessed a series of lights in a V-shaped formation over Phoenix, Arizona.

- The Event: The lights appeared in the sky over a 300-mile area from Nevada to Phoenix. Witnesses described a massive, silent craft with five spherical lights or a series of individual lights moving in formation. The phenomenon lasted for several hours and was observed by people of all ages and backgrounds, including the Governor of Arizona at the time, Fife Symington.

- Explanations: The most common explanation offered by the military was that the lights were flares dropped during an Air National Guard training exercise at the Barry Goldwater Range. However, many witnesses and researchers dispute this explanation, arguing that the lights' behavior and appearance do not match that of military flares.

- Legacy: The Phoenix Lights remain one of the most well-documented and controversial UFO sightings. The event has been the subject of numerous documentaries, books, and investigations, and continues to be a key case study for those interested in UFO phenomena.

"Time travel used to be thought of as just science fiction, but Einstein's general theory of relativity allows for the possibility that we could warp spacetime so much that you could return to the past."

— Stephen Hawking

Government Investigations:

Over the years, various governments, particularly the United States, have conducted investigations into UFO phenomena.

- Project Blue Book (1952-1969): Conducted by the U.S. Air Force, this was one of the most extensive investigations into UFO sightings. It aimed to determine if UFOs were a threat to national security and to scientifically analyze UFO-related data. Out of 12,618 sightings, 701 remain "unidentified."

- The Condon Report (1968): Commissioned by the U.S. Air Force and conducted by the University of Colorado, this report concluded that UFOs were not a threat and that further scientific study of UFO phenomena was unlikely to yield significant results. The findings led to the termination of Project Blue Book.

- Advanced Aerospace Threat Identification Program (AATIP) (2007-2012): This secretive Pentagon program aimed to study UFOs, specifically those that posed a potential threat to national security. Details of the program and some of its findings were disclosed to the public in 2017, reigniting interest and debate over UFO phenomena.

Recent Government Disclosures

In recent years, there has been a significant shift towards transparency regarding UFOs, with several notable disclosures and reports.

- The 2020 Pentagon Videos: In April 2020, the Pentagon officially released three videos recorded by Navy pilots that showed unidentified aerial phenomena (UAPs). These videos, previously leaked, depict objects moving at high speeds with no visible means of propulsion, raising questions about their origin.

- The UAP Task Force (2020): Established by the Department of Defense, this task force aims to improve the understanding of and gain insight into the nature and origins of UAPs. The task force's findings are expected to inform the development of new policies and procedures for military training and operations.

- The 2021 Director of National Intelligence Report: This report, released in June 2021, reviewed 144 UAP incidents reported by U.S. government sources since 2004. The report concluded that most of the incidents could not be explained and recommended further scientific investigation.

Public and Scientific Interest

The recent government disclosures have spurred renewed interest from both the public and the scientific community.

- Public Fascination: UFO phenomena have long captured the public's imagination, inspiring countless books, movies, and television shows. Recent disclosures have only heightened interest, leading to increased demand for transparency and further investigation.

- Scientific Investigation: With the stigma surrounding UFO research gradually diminishing, more scientists are advocating for rigorous study of UAPs. Organizations like the Scientific Coalition for UAP Studies (SCU) are dedicated to the scientific examination of UAP phenomena, aiming to bring credibility and data-driven analysis to the field.

"The past is a foreign country: they do things differently there."

— *L.P. Hartley*

Synthetic Biology: Combining Biology and Engineering

Synthetic biology is an interdisciplinary field that merges biology and engineering to design and construct new biological parts, devices, and systems. It also involves the re-design of existing natural biological systems for useful purposes. By applying engineering principles to biology, scientists can create organisms with novel functions, potentially revolutionizing various sectors such as medicine, agriculture, and environmental science.

Core Concepts and Techniques

- Genetic Engineering: At the heart of synthetic biology is the manipulation of an organism's genetic material. Techniques such as CRISPR-Cas9 allow scientists to precisely edit genes, enabling the creation of organisms with desired traits. For example, bacteria can be engineered to produce insulin, a critical hormone for diabetes treatment.

- Standardized Biological Parts: Synthetic biologists use standardized DNA sequences, known as BioBricks, which can be combined in various ways to build new genetic constructs. These BioBricks function like interchangeable parts in engineering, allowing for more predictable and modular designs.

- Pathway Engineering: This involves the reconfiguration of metabolic pathways within an organism to produce new substances. For example, yeast can be engineered to produce biofuels or pharmaceuticals by introducing new metabolic pathways.

- Artificial Life: Researchers are exploring the creation of entirely synthetic life forms. In 2010, the J. Craig Venter Institute created the first synthetic bacterial cell by synthesizing a complete genome and transplanting it into a host cell.

Applications of Synthetic Biology

- Medicine: Synthetic biology holds immense potential in the medical field. Engineered bacteria can be used as "living medicines" to treat infections, while synthetic biology techniques are being employed to develop new vaccines and antibiotics. Additionally, gene therapy approaches are being enhanced to treat genetic disorders more effectively.

- Agriculture: Crops can be genetically modified for improved yield, resistance to pests, and tolerance to environmental stresses. Synthetic biology also enables the development of biofertilizers and biopesticides that are more sustainable and eco-friendly compared to traditional chemicals.

- Environmental Sustainability: Synthetic organisms can be designed to degrade pollutants, clean up oil spills, or capture carbon dioxide from the atmosphere. This makes synthetic biology a powerful tool in combating environmental issues and promoting sustainability.

- Biofuels and Bioproducts: Engineered microorganisms can convert biomass into biofuels, providing a renewable alternative to fossil fuels. Synthetic biology also enables the production of bioplastics and other biodegradable materials, reducing reliance on petroleum-based products.

Ethical and Safety Considerations

While synthetic biology offers remarkable opportunities, it also raises significant ethical and safety concerns. One major issue is biosecurity: the creation of new organisms poses potential risks if they are accidentally released into the environment or used maliciously.

Ensuring robust containment measures and developing strict ethical guidelines are crucial to prevent unintended consequences. Moreover, the ability to create and modify life forms raises profound ethical questions about the extent to which humans should interfere with natural processes. Addressing these concerns requires extensive public dialogue and the establishment of comprehensive regulatory frameworks.

Another critical consideration is intellectual property. The patenting of synthetic biological parts and organisms could restrict access to essential technologies, potentially impeding scientific progress. Balancing innovation with open access is a key challenge that must be navigated carefully. Ensuring that the benefits of synthetic biology are widely accessible while protecting the rights of innovators will require thoughtful policies and collaboration across sectors. By addressing these ethical and safety considerations, society can responsibly harness the potential of synthetic biology for the greater good.

The Future of Synthetic Biology

The future of synthetic biology is promising, with rapid advancements and growing interdisciplinary collaboration. As technology progresses, we can expect more sophisticated synthetic organisms and applications that could transform industries and improve human life. Continued research, responsible innovation, and public engagement will be vital in harnessing the full potential of synthetic biology.

Exoplanets and the Search for Extraterrestrial Life: Fact File

First Exoplanet Discovery: The first confirmed exoplanet, 51 Pegasi b, was discovered in 1995, orbiting a Sun-like star 50 light-years away.

Number of Exoplanets: As of 2024, over 5,000 exoplanets have been confirmed in our galaxy, the Milky Way.

Goldilocks Zone: Exoplanets located in the "habitable zone" or "Goldilocks zone" of their stars have conditions that might support liquid water, an essential ingredient for life as we know it.

Kepler Space Telescope: Launched in 2009, the Kepler Space Telescope has discovered more than 2,700 exoplanets, significantly advancing our knowledge of these distant worlds.

Transit Method: The most successful method for discovering exoplanets is the transit method, which detects planets by observing the slight dimming of a star as an orbiting planet passes in front of it.

Radial Velocity Method: This method detects exoplanets by measuring the star's wobble caused by the gravitational pull of an orbiting planet, revealing its presence and mass.

Proxima Centauri b: The closest known exoplanet to Earth, Proxima Centauri b, orbits the star Proxima Centauri just 4.24 light-years away and lies within its star's habitable zone.

Rogue Planets: Some exoplanets, known as rogue planets, do not orbit any star and instead drift through space.

Water Worlds: Scientists believe that some exoplanets, called "water worlds," may have surfaces entirely covered by deep oceans, potentially harboring life.

Super-Earths: Exoplanets larger than Earth but smaller than Neptune are known as super-Earths. They could have conditions suitable for life and are common in the galaxy.

Atmospheric Analysis: By studying the atmospheres of exoplanets using spectroscopy, scientists can detect potential biosignatures like oxygen, methane, and water vapor that might indicate the presence of life.

Exoplanet Diversity: Exoplanets come in various types, including gas giants, rocky planets, ice giants, and even lava worlds with molten surfaces.

TRAPPIST-1 System: The TRAPPIST-1 system has seven Earth-sized exoplanets, three of which lie in the habitable zone, making it a prime target in the search for extraterrestrial life.

James Webb Space Telescope: The James Webb Space Telescope, launched in 2021, is capable of analyzing the atmospheres of exoplanets with unprecedented detail, advancing the search for life beyond Earth.

SETI (Search for Extraterrestrial Intelligence): The SETI program uses radio telescopes to listen for signals from intelligent alien civilizations. While no definitive signals have been found yet, the search continues.

Psychokinesis: Mind-Over-Matter Phenomena

What is Psychokinesis?

Psychokinesis, also known as telekinesis, is the purported ability of the mind to influence physical objects without physical interaction. This phenomenon suggests that individuals can move objects, bend spoons, or alter electronic devices using only their mental power. While it has been a popular subject in folklore, fiction, and parapsychology, psychokinesis remains highly controversial and is not widely accepted by the scientific community due to the lack of empirical evidence.

Why is Psychokinesis Studied?

The study of psychokinesis attracts interest for several reasons:

1. Understanding Human Potential: Investigating psychokinesis could potentially unlock unknown aspects of human cognition and capability.

2. Scientific Curiosity: Scientists are driven by curiosity to explore and explain phenomena that challenge conventional understanding.

3. Paranormal Research: Psychokinesis is a key interest within the field of parapsychology, which seeks to investigate and understand paranormal experiences and abilities.

4. Cultural Impact: Psychokinesis has significant cultural and historical relevance, often featuring in stories, films, and folklore.

15 Crazy Facts About Psychokinesis

1. Uri Geller: Uri Geller is one of the most famous individuals associated with psychokinesis, claiming to bend spoons and keys with his mind. His feats have been both celebrated and heavily scrutinized by skeptics.

2. The Philip Experiment (1972): In Toronto, a group of researchers created a fictional ghost named Philip and, through collective mental focus, claimed to produce psychokinetic effects such as table movements.

3. The Poltergeist Connection: Many reported poltergeist cases involve unexplained psychokinetic activities like objects moving or flying across rooms, often centered around a particular individual, usually an adolescent.

4. Laboratory Research: Parapsychologist J.B. Rhine conducted early laboratory experiments on psychokinesis at Duke University, using dice-throwing experiments to test if participants could influence outcomes.

5. The PEAR Lab: The Princeton Engineering Anomalies Research (PEAR) laboratory conducted experiments on psychokinesis for nearly 30 years, exploring if human intention could affect random event generators.

6. Micro-PK: Micro-psychokinesis refers to the supposed ability to influence very small physical systems, like atomic or quantum systems, which is a focus in many laboratory experiments.

7. Skeptics' Challenge: Famous skeptic James Randi offered a $1 million prize for anyone who could demonstrate psychokinetic abilities under controlled conditions. The prize remains unclaimed.

8. Electronic Interference: Some individuals claim to have the ability to affect electronic devices, such as making watches stop or causing lights to flicker, purely through mental focus.

9. Historical Accounts: Historical figures like Daniel Dunglas Home were famous for their alleged psychokinetic abilities, performing levitations and moving objects without touching them.

10. Cultural Representations: Psychokinesis is a popular theme in movies and TV shows, with characters like Eleven from "Stranger Things" and the Jedi from "Star Wars" showcasing dramatic psychokinetic abilities.

11. Quantum Mechanics Hypotheses: Some speculative theories suggest that quantum mechanics might provide a framework for understanding psychokinesis, though these ideas are far from scientifically validated.

12. Spoon Bending Parties: In the 1970s, "spoon bending parties" became a fad, where groups of people would gather and attempt to bend spoons using psychokinesis, often led by individuals claiming special abilities.

13. Online Communities: There are numerous online communities and forums where enthusiasts share tips, techniques, and personal experiences related to psychokinesis.

14. Controlled Experiments: Strictly controlled scientific experiments generally have not found conclusive evidence to support psychokinesis, leading to ongoing debate between believers and skeptics.

15. Military Interest: There have been rumors and speculative reports about military interest in psychokinesis for applications in espionage and warfare, though concrete evidence is lacking.

Near-Death Experiences (NDEs): Understanding Consciousness

Near-death experiences (NDEs) are profound psychological events that occur to individuals who are close to death or have been in situations where death was imminent. These experiences often involve a range of sensations, including feelings of peace, detachment from the body, seeing a bright light, and encounters with deceased loved ones or spiritual beings. NDEs have been reported across cultures and throughout history, prompting extensive research to understand their nature and implications.

The study of NDEs is important for several reasons:

1. **Exploring Consciousness:** NDEs provide a unique opportunity to investigate the nature of human consciousness and what happens when the brain is near death.
2. **Medical Implications:** Understanding NDEs can help healthcare professionals support patients who have experienced them and integrate these experiences into their recovery process.
3. **Psychological Insights:** NDEs can offer insights into the psychological processes involved in life-threatening situations and their impact on individuals' mental health.
4. **Spiritual and Philosophical Questions:** NDEs raise important questions about life, death, and the possibility of an afterlife, influencing religious and philosophical perspectives.

"Many people who have had near-death experiences say that the most important lesson they learned was how to love."

— Dr. Raymond Moody

Crazy Facts About Near-Death Experiences

1. **Common Features:** Despite cultural differences, many NDEs share common elements such as out-of-body experiences, tunnel vision, and encounters with a bright light or beings of light.
2. **Life Review:** Many individuals report experiencing a life review, where they relive significant moments of their lives, often with a focus on the emotional impact of their actions on others.
3. **Heightened Senses:** During an NDE, people often describe having heightened senses and an acute awareness, even if their physical body is unresponsive.
4. **Transformed Lives:** Many people who have had NDEs report profound changes in their attitudes, values, and outlook on life, often becoming more spiritual and less afraid of death.
5. **Cardiac Arrest Studies:** Research involving cardiac arrest patients shows that a significant percentage report NDEs, providing a substantial population for scientific study.
6. **The Pam Reynolds Case:** One of the most famous NDE cases involves Pam Reynolds, who reported detailed experiences during a surgical procedure in which her brain activity was completely flatlined.
7. **NDEs in Children:** Children also report NDEs, often with similar elements to those experienced by adults, suggesting these experiences are not purely culturally learned.
8. **Scientific Skepticism:** Some scientists argue that NDEs can be explained by physiological and neurological processes, such as lack of oxygen to the brain, release of endorphins, or temporal lobe seizures.
9. **Cultural Variations:** The content of NDEs can vary significantly across cultures, with individuals often experiencing elements consistent with their cultural and religious beliefs.

10. **Blind Individuals:** There are documented cases of blind individuals reporting visual experiences during NDEs, which they claim are their first-ever visual perceptions.
11. **The AWARE Study:** The AWARE (AWAreness during REsuscitation) study, led by Dr. Sam Parnia, investigates the brain activity and consciousness of patients who have undergone cardiac arrest, aiming to scientifically understand NDEs.
12. **Lucid Death:** Some researchers propose the concept of "lucid death," where the mind remains aware even as the body shuts down, as a potential explanation for NDEs.
13. **Neurochemical Theories:** Hypotheses involving neurotransmitters like serotonin and endorphins suggest that these chemicals may play a role in the vivid and profound nature of NDEs.
14. **Shared Death Experiences:** Similar to NDEs, shared death experiences occur when bystanders feel they accompany the dying person partway into the afterlife, experiencing similar phenomena.
15. **NDE Research Organizations:** Organizations like the International Association for Near-Death Studies (IANDS) and the Near-Death Experience Research Foundation (NDERF) are dedicated to studying and understanding NDEs, offering support and information to experiencers.

Anita Moorjani:

"After my near-death experience, I realized that life is a gift, and it's up to us to find the beauty and joy in each moment."
— Anita Moorjani

Harry Houdini: The Famous Magician and Skeptic

Harry Houdini, born Erik Weisz on March 24, 1874, in Budapest, Hungary, immigrated with his family to the United States at the age of four, settling in Appleton, Wisconsin. He chose his stage name in honor of Jean-Eugène Robert-Houdin, a famous French magician, believing that adding an 'i' to Houdin's name meant "like Houdin" in French. Houdini's early career was marked by struggle as he performed in dime museums and sideshows, gaining little success. His breakthrough came in 1899 when manager Martin Beck booked him on a successful tour of vaudeville theaters, launching his career as the "Handcuff King" known for escaping police handcuffs and other restraints.

Houdini earned his reputation as a master of escapology, performing daring escapes from straitjackets, water-filled milk cans, and the infamous Chinese Water Torture Cell. Beyond his incredible performances, Houdini became a fervent skeptic of spiritualism, a belief that the dead can communicate with the living. Following the death of his mother, he embarked on a personal mission to expose fraudulent spiritualists who preyed on grieving families. Using his knowledge of magic, Houdini attended séances in disguise and revealed the tricks spiritualists used to fake paranormal phenomena.

Initially, Houdini was friends with Sir Arthur Conan Doyle, the creator of Sherlock Holmes and a strong believer in spiritualism. However, their friendship soured as Houdini's public debunking of spiritualists clashed with Doyle's firm belief in the supernatural. Houdini wrote several books and gave numerous lectures debunking spiritualism, including "A Magician Among the Spirits" (1924), and his efforts laid the groundwork for future skeptics and debunkers, including organizations like the Committee for Skeptical Inquiry.

Houdini's life came to an end on October 31, 1926, after he suffered from peritonitis caused by a ruptured appendix. Some believe his death was hastened by a punch to the stomach he received days earlier, a demonstration he often performed to show his physical endurance. Before his death, Houdini promised his wife Bess that he would try to communicate from beyond the grave if possible. Every Halloween, a séance is held to attempt contact with Houdini's spirit, though no definitive contact has ever been made.

Houdini's legacy continues to inspire both the world of magic and the skeptical movement. His life and exploits have been the subject of numerous films, TV shows, and documentaries. He remains an iconic figure for his mastery in magic and his dedication to debunking fraudulent spiritualists.

"My chief task has been to conquer fear. The public sees only the thrill of the escape, never the details of the private fight which has taken place to win the honor of making the attempt."

Extra Tidbits:

- **Overboard Box Escape:** In 1912, Houdini was shackled, placed in a wooden crate, and lowered into New York's East River, escaping within minutes.
- **Suspended Straitjacket Escape:** Houdini frequently performed this escape, hanging upside down in a straitjacket from skyscrapers, drawing large crowds.
- **Secret Code with Bess:** Houdini and his wife Bess agreed on a secret code to confirm any genuine communication from beyond the grave. The code has never been correctly revealed by any medium.

- **Magician's Oath:** Houdini was a firm believer in the Magician's Oath, which forbids magicians from revealing their secrets to the public.
- **Library Collection:** The Library of Congress holds the largest collection of Houdini's personal papers, photographs, and memorabilia, donated by his family.
- **Enduring Influence:** Houdini's legacy as a master magician and a pioneering skeptic endures, inspiring countless magicians, escape artists, and skeptics worldwide.
- **Metamorphosis Act:** Houdini and his wife Bess performed a famous illusion called "Metamorphosis," where they would switch places in a locked trunk within seconds.
- **Underwater Buried Alive:** Houdini performed a dangerous stunt where he was buried alive and had to escape from a coffin six feet underground.
- **Film Career:** Houdini starred in several silent films, including "The Master Mystery" and "The Grim Game," using his skills to perform daring stunts on screen.
- **Handcuff Secrets:** Houdini collected handcuffs from police forces around the world and studied their mechanisms, allowing him to escape from nearly any pair.
- **Escaping from Jails:** Houdini would often challenge local jails to lock him up, and he would escape to demonstrate his skills and promote his shows.
- **Death-Defying Publicity Stunts:** To draw crowds and generate publicity, Houdini frequently performed dangerous public stunts, such as escaping from a straitjacket while suspended from a crane over a busy street.

"What the eyes see and the ears hear, the mind believes."

Nikola Tesla: The Genius Inventor

Nikola Tesla was a Serbian-American inventor, electrical engineer, mechanical engineer, and physicist, best known for his groundbreaking work in developing the modern alternating current (AC) electricity supply system. Born on July 10, 1856, in Smiljan, in what is now Croatia, Tesla showed an early aptitude for mathematics and engineering. He moved to the United States in 1884, where he briefly worked with Thomas Edison before striking out on his own. Tesla's inventions and theoretical work formed the basis of modern AC power systems, including the Tesla coil, which is still used in radio technology today. His visionary ideas extended beyond electricity; he envisioned wireless communication, remote control, and even concepts for renewable energy.

Despite his incredible contributions to science and technology, Tesla struggled financially and was often overshadowed by contemporaries like Edison. He spent his later years in relative obscurity, dying in New York City on January 7, 1943. Today, Tesla is celebrated as a visionary who was ahead of his time, and his legacy lives on in the numerous technologies that his work helped to pioneer.

Here are ten **true or false** statements about Nikola Tesla to test your knowledge:

1. Nikola Tesla had a photographic memory and was known to memorize entire books and images.

2. Tesla developed the idea for a particle beam weapon, which he referred to as the "Teleforce" or "Death Ray."

3. Nikola Tesla was fluent in eight languages, including Serbian, English, French, German, and Italian.

4. Tesla once worked as a ditch digger for $2 per day to make ends meet after leaving Edison's company.

5. Nikola Tesla's design for an induction motor was patented and played a crucial role in the development of AC power.

6. Tesla experimented with cryogenic technology and believed it could significantly increase the lifespan of humans.

7. Nikola Tesla received the Edison Medal in 1917, which was the most prestigious award given by the American Institute of Electrical Engineers.

8. Tesla's Wardenclyffe Tower was intended to provide free wireless electricity to the entire world.

9. Tesla claimed to have developed a technique for photographing thoughts, believing that images formed in the mind could be captured.

10. Tesla was obsessed with the number three and performed many of his daily routines in sets of three.

Answers - True or False

True: Nikola Tesla had a photographic memory and was known to memorize entire books and images.

True: Tesla developed the idea for a particle beam weapon, which he referred to as the "Teleforce" or "Death Ray."

True: Nikola Tesla was fluent in eight languages, including Serbian, English, French, German, and Italian.

True: Tesla once worked as a ditch digger for $2 per day to make ends meet after leaving Edison's company.

True: Nikola Tesla's design for an induction motor was patented and played a crucial role in the development of AC power.

False: Tesla did not experiment with cryogenic technology nor did he believe it could significantly increase the lifespan of humans.

True: Nikola Tesla received the Edison Medal in 1917, which was the most prestigious award given by the American Institute of Electrical Engineers.

True: Tesla's Wardenclyffe Tower was intended to provide free wireless electricity to the entire world.

False: Tesla did not develop a technique for photographing thoughts, though he did have many visionary ideas that extended beyond his time.

True: Tesla was obsessed with the number three and performed many of his daily routines in sets of three.

"One must be sane to think clearly, but one can think deeply and be quite insane." — Nikola Tesla

Why Elon Musk Chose Nikola Tesla's Name for His Company

Elon Musk named his electric vehicle company "Tesla" as a tribute to Nikola Tesla, the visionary inventor who pioneered many foundational technologies in electricity and electromagnetism. Tesla's work on alternating current (AC) power systems revolutionized the way electricity is generated and distributed, making long-distance transmission of electrical power feasible. By choosing Tesla's name, Musk aimed to honor the legacy of a man who had a profound impact on modern electrical engineering and who symbolized innovation and forward-thinking technology. Musk's vision for his company mirrored Tesla's aspirations for a world powered by clean, sustainable energy sources.

Moreover, Tesla's reputation as a futurist and his dedication to improving humanity through technology align closely with the mission of Musk's companies, including SpaceX and Tesla, Inc. Nikola Tesla's dream of wireless energy transmission and his numerous other inventions reflected a relentless pursuit of progress and a desire to push the boundaries of what was considered possible. By naming the company after Nikola Tesla, Elon Musk not only acknowledges the historical significance of Tesla's work but also sets a high standard for innovation, creativity, and transformative impact in the fields of energy and transportation.

"The present is theirs; the future, for which I really worked, is mine." — Nikola Tesla

Sally Ride: The First American Woman in Space

1. Early Life:

Birth: Sally Kristen Ride was born on May 26, 1951, in Encino, California.

Education: Ride attended Stanford University, where she earned a Bachelor's degree in English and Physics in 1973. She continued her studies at Stanford, earning a Master's degree and a Ph.D. in Physics.

2. NASA Career:

Selection: In 1978, Sally Ride was selected as one of the first six women to join NASA's astronaut program, out of a class of 35 astronaut candidates.

Historic Flight: On June 18, 1983, Ride made history by becoming the first American woman in space aboard the Space Shuttle Challenger (STS-7). At 32 years old, she was also the youngest American astronaut to have flown in space at that time.

Second Mission: Ride flew on a second space mission in 1984, also aboard Challenger (STS-41-G). During her two missions, she spent a total of over 343 hours in space.

3. Contributions and Legacy:

Role in Challenger Investigation: After the Challenger disaster in 1986, Ride was appointed to the Presidential Commission investigating the accident, contributing her expertise to the inquiry.

NASA's Office of Exploration: She later served at NASA Headquarters, where she helped develop strategic plans for the future of space exploration.

Sally Ride Science: After retiring from NASA, Ride co-founded Sally Ride Science in 2001, an organization dedicated to inspiring young people, particularly girls, to pursue careers in science, technology, engineering, and math (STEM).

4. Awards and Honors:

Inductions: Ride was inducted into the National Women's Hall of Fame and the Astronaut Hall of Fame.

Medals: She received numerous awards, including the NASA Space Flight Medal and the NCAA's Theodore Roosevelt Award.

Posthumous Honors: After her death, President Barack Obama awarded her the Presidential Medal of Freedom in 2013.

5. Personal Life:

Privacy: Ride was known for her privacy. She revealed few details about her personal life, including her 27-year relationship with Tam O'Shaughnessy, her partner and co-founder of Sally Ride Science, which became public only after her death.

6. Advocacy for Education:

Books and Outreach: Ride authored several children's books on space and science to encourage young readers to engage with these subjects.

Legacy in Education: Her work with Sally Ride Science emphasized the importance of hands-on learning and the inclusion of girls and minorities in STEM fields.

7. Death:

Passing: Sally Ride passed away on July 23, 2012, at the age of 61, after a 17-month battle with pancreatic cancer.

Legacy: Ride's legacy continues through her contributions to science, her inspiration to future generations of astronauts, and her advocacy for STEM education.

Extra Tidbits

- Pioneer for Women: Sally Ride broke significant barriers for women in space and became a role model for aspiring female astronauts and scientists worldwide.

- Space Shuttle Missions: During her missions, Ride operated the shuttle's robotic arm to deploy and retrieve satellites, showcasing her technical skills.
- First Ride of the Challenger: Ride's first mission aboard the Challenger was only the seventh space shuttle mission in NASA's history.
- Academic Contributions: Besides her work at NASA, Ride also served as a professor of physics at the University of California, San Diego (UCSD).
- Sally Ride EarthKAM: She initiated the EarthKAM project, allowing middle school students to take pictures of Earth from space and learn about geography, environmental science, and space technology.

"Young girls need to see role models in whatever careers they may choose, just so they can picture themselves doing those jobs someday. You can't be what you can't see."

— Sally Ride

Interaction of Science and Geography

Geography and science are deeply intertwined in the quest to understand the Earth and its complex processes. Physical geography focuses on the natural environment, encompassing various scientific disciplines such as geology, meteorology, hydrology, and biology. These sciences help geographers analyze the physical features of the Earth, including mountains, rivers, climate patterns, and ecosystems. For instance, meteorologists study weather and climate, providing essential data that geographers use to understand regional climates and their impacts on human activities and natural landscapes. Similarly, geology informs geography by explaining the formation and structure of the Earth, enabling geographers to map out and interpret landforms and natural resources.

Human geography examines the relationships between people and their environments, integrating insights from social sciences and natural sciences. Environmental science, a field that bridges geography and science, studies how natural and human systems interact, focusing on issues such as pollution, resource management, and sustainability. Geographic Information Systems (GIS) are a prime example of this integration, utilizing scientific data to analyze spatial patterns and processes. GIS technology allows geographers to create detailed maps and models that inform urban planning, disaster management, and environmental conservation efforts. By combining scientific research with geographic analysis, scientists and geographers can address global challenges such as climate change, deforestation, and urbanization, providing critical insights for policy-making and sustainable development.

- The theory of plate tectonics, a fundamental concept in geology, explains the movement of Earth's lithospheric plates. This movement causes earthquakes, volcanic activity, and the formation of mountains, which are key topics in physical geography.
- Meteorology and climatology provide the basis for understanding climate zones, which in turn define the world's biomes. These biomes, such as deserts, forests, and tundras, are mapped and studied in geography for their ecological and human significance.
- Hydrology studies the distribution and movement of water on Earth, which is crucial for understanding the hydrological cycle. Geography uses this scientific knowledge to analyze water resources, river systems, and the impact of water on landscapes and human settlements.
- Advances in remote sensing, a technology used in both science and geography, allow scientists to collect data about Earth's surface from satellites. This data helps in mapping land use, monitoring environmental changes, and managing natural resources.
- Scientific studies of natural disasters such as hurricanes, floods, and tsunamis inform geographic strategies for disaster risk management and mitigation. Geographers use scientific data to plan and implement measures to reduce the impact of these events on human populations.
- Ecology, a branch of biology, studies ecosystems and their functions. Geographers use this scientific understanding to evaluate ecosystem services, such as pollination, water purification, and carbon sequestration, which are vital for human well-being and environmental sustainability.

Gravitational Waves

Gravitational waves are ripples in the fabric of spacetime, predicted by Albert Einstein's General Theory of Relativity in 1915. These waves are generated by the acceleration of massive objects, such as the collision and merger of black holes or neutron stars. For decades, scientists considered the detection of gravitational waves a formidable challenge due to the incredibly small disturbances they cause. It wasn't until the advancement of highly sensitive instrumentation that the direct observation of these waves became possible, leading to groundbreaking discoveries that have opened a new era in astrophysics.

The first direct detection of gravitational waves was announced on February 11, 2016, by the LIGO (Laser Interferometer Gravitational-Wave Observatory) collaboration. This momentous event confirmed the merging of two black holes approximately 1.3 billion light-years away, producing a powerful burst of gravitational waves. The signal, named GW150914, matched the predictions of Einstein's equations, providing compelling evidence for the existence of these elusive waves. The discovery earned the 2017 Nobel Prize in Physics and validated the use of gravitational waves as a new tool for observing the cosmos.

Gravitational waves have profound implications for our understanding of the universe. They offer a unique method of probing astronomical phenomena that are otherwise invisible, such as black holes and neutron star mergers. Unlike electromagnetic waves, which can be absorbed or scattered by matter, gravitational waves pass through matter almost unimpeded, carrying information about their sources. This allows scientists to study the properties and dynamics of some of the most extreme environments in the universe, offering insights into fundamental physics and the behavior of matter under extreme conditions.

The detection of gravitational waves also provides a new way to test the limits of General Relativity and our understanding of gravity. By comparing the observed waveforms with theoretical predictions, physicists can investigate the validity of Einstein's theory in strong gravitational fields. Any deviations could hint at new physics beyond the Standard Model, potentially leading to a deeper understanding of the fundamental forces and particles that make up our universe. This line of research could also shed light on the mysterious nature of dark matter and dark energy, which together constitute about 95% of the universe's mass-energy content.

Looking to the future, the field of gravitational wave astronomy is poised for rapid expansion. Planned upgrades to existing detectors, such as LIGO and Virgo, along with new observatories like KAGRA in Japan and the upcoming space-based LISA (Laser Interferometer Space Antenna), will enhance our ability to detect and analyze gravitational waves. These advancements will increase the sensitivity and frequency range of observations, allowing scientists to explore a broader array of cosmic events. As our observational capabilities grow, gravitational waves will continue to revolutionize our understanding of the universe, revealing previously hidden aspects of its structure and evolution.

Did You Know?

The first detected gravitational wave event, GW150914, involved the collision of two black holes with masses of approximately 29 and 36 times that of the Sun. During their final moments, they released an amount of energy equivalent to about three solar masses, all within a fraction of a second. This energy output was greater than the combined power of all the stars in the observable universe at that moment, illustrating the immense forces at play during such cosmic events.

Thank You for Reading!

Dear Readers,

Thank you for joining us on this incredible journey through the realms of science, space, and supernatural phenomena. We hope you found this edition of Intelligent Minds as stimulating and fascinating as we did in creating it. Your curiosity and enthusiasm make this exploration truly special.

As you turned each page, we aimed to challenge your understanding, pique your interest, and inspire you to delve deeper into these intriguing subjects. Whether you are a seasoned scientist, aspiring astronomer, or curious skeptic, we trust that the content has expanded your perspective and fueled your passion for discovery.

We are delighted to offer you even more opportunities to learn and explore. Scan the QR code below to access free online content, including additional trivia, interactive quizzes, in-depth articles, and exclusive interviews with experts. This digital extension of our book provides endless avenues for expanding your knowledge and satisfying your curiosity.

Thank you once again for your time and interest. We look forward to continuing this journey of discovery with you in future editions of Intelligent Minds. Stay curious, keep exploring, and never stop questioning the world around you.

Warm regards,

BNW Team

Printed in Great Britain
by Amazon